Data Analysis Using the Method of Least Squares

J. Wolberg

Data Analysis Using the Method of Least Squares

Extracting the Most Information from Experiments

With 58 Figures and 68 Tables

 Springer

John Wolberg
Technion-Israel Institute of Technology
Faculty of Mechanical Engineering
32000 Haifa, Israel
E-mail: jwolber@attglobal.net

Library of Congress Control Number: 2005934230

ISBN 978-3-540-25674-8 Springer Berlin Heidelberg New York

Springer is a part of Springer Science+Business Media.

springer.com

© Springer-Verlag Berlin Heidelberg 2006

Typesetting: Data prepared by the Author and by SPI Publisher Services
Cover design: *design & production* GmbH, Heidelberg

Printed on acid-free paper   SPIN 11010197   62/3141/SPI Publisher Services   5 4 3 2 1 0

For my parents, Sidney and Beatrice Wolberg ז״ל

My wife Laurie

My children and their families:

Beth, Gilad, Yoni and Maya Sassoon
David, Pazit and Sheli Wolberg
Danny, Iris, Noa, Adi and Liat Wolberg
Tamar, Ronen, Avigail and Aviv Kimchi

Preface

Measurements through quantitative experiments are one of the most fundamental tasks in all areas of science and technology. Astronomers analyze data from asteroid sightings to predict orbits. Computer scientists develop models for recognizing spam mail. Physicists measure properties of materials at low temperatures to understand superconductivity. Materials engineers study the reaction of materials to varying load levels to develop methods for prediction of failure. Chemical engineers consider reactions as functions of temperature and pressure. The list is endless. From the very small-scale work on DNA to the huge-scale study of black holes, quantitative experiments are performed and the data must be analyzed.

Probably the most popular method of analysis of the data associated with quantitative experiments is least squares. It has been said that the method of least squares was to statistics what calculus was to mathematics. Although the method is hardly mentioned in most engineering and science undergraduate curricula, many graduate students end up using the method to analyze the data gathered as part of their research. There is not a lot of available literature on the subject. Very few books deal with least squares at the level of detail that the subject deserves. Many books on statistics include a chapter on least squares but the treatment is usually limited to the simplest cases of linear least squares. The purpose of this book is to fill the gaps and include the type of information helpful to scientists and engineers interested in applying the method in their own special fields.

The purpose of many engineering and scientific experiments is to determine parameters based upon a mathematical model related to the phenomenon under observation. Even if the data is analyzed using least squares, the full power of the method is often overlooked. For example, the data can be weighted based upon the estimated errors associated with the data. Results from previous experiments or calculations can be combined with the least squares analysis to obtain improved estimate of the model parameters. In addition, the results can be used for predicting values of the dependent variable or variables and the associated uncertainties of the predictions as functions of the independent variables.

The introductory chapter (Chapter 1) includes a review of the basic statistical concepts that are used throughout the book. The method of least squares is developed in Chapter 2. The treatment includes development of mathematical models using both linear and nonlinear least squares. In Chapter 3 evaluation of models is considered. This chapter includes methods for measuring the "goodness of fit" of a model and methods for comparing different models. The subject of candidate predictors is discussed in Chapter 4. Often there are a number of candidate predictors and the task of the analyst is to try to extract a model using subspaces of the full candidate predictor space. In Chapter 5 attention is turned towards designing experiments that will eventually be analyzed using least squares. The subject considered in Chapter 6 is nonlinear least squares software. Kernel regression is introduced in the final chapter (Chapter 7). Kernel regression is a nonparametric modeling technique that utilizes local least squares estimates.

Although general purpose least squares software is available, the subject of least squares is simple enough so that many users of the method prefer to write their own routines. Often, the least squares analysis is a part of a larger program and it is useful to imbed it within the framework of the larger program. Throughout the book very simple examples are included so that the reader can test his or her own understanding of the subject. These examples are particularly useful for testing computer routines.

The REGRESS program has been used throughout the book as the primary least squares analysis tool. REGRESS is a general purpose nonlinear least squares program and I am its author. The program can be downloaded from *www.technion.ac.il/wolberg*.

I would like to thank David Aronson for the many discussions we have had over the years regarding the subject of data modeling. My first experiences with the development of general purpose nonlinear regression software were influenced by numerous conversations that I had with Marshall Rafal. Although a number of years have passed, I still am in contact with Marshall. Most of the examples included in the book were based upon software that I developed with Ronen Kimchi and Victor Leikehman and I would like to thank them for their advice and help. I would like to thank Ellad Tadmor for getting me involved in the research described in Section 7.7. Thanks to Richard Green for introducing me to the first English translation of Gauss's *Theoria Motus* in which Gauss developed the foundations of the method of least squares. I would also like to thank Donna Bossin for her help in editing the manuscript and teaching me some of the cryptic subtleties of WORD.

I have been teaching a graduate course on analysis and design of experiments and as a result have had many useful discussions with our students throughout the years. When I decided to write this book two years ago, I asked each student in the course to critically review a section in each chapter that had been written up to that point. Over 20 students in the spring of 2004 and over 20 students in the spring of 2005 submitted reviews that included many useful comments and ideas. A number of typos and errors were located as a result of their efforts and I really appreciated their help.

John R. Wolberg
Haifa, Israel
July, 2005

Contents

Chapter 1 INTRODUCTION

1.1 Quantitative Experiments

Most areas of science and engineering utilize **quantitative experiments** to determine parameters of interest. Quantitative experiments are characterized by measured variables, a mathematical model and unknown parameters. For most experiments the method of **least squares** is used to analyze the data in order to determine values for the unknown parameters.

As an example of a quantitative experiment, consider the following: measurement of the half-life of a radioactive isotope. Half-life is defined as the time required for the count rate of the isotope to decrease by one half. The experimental setup is shown in Figure 1.1.1. Measurements of *Counts* (i.e., the number of counts observed per time unit) are collected from time 0 to time *tmax*. The mathematical model for this experiment is:

$$Counts = amplitude \cdot e^{-decay\_constant \cdot t} + background \qquad (1.1.1)$$

For this experiment, *Counts* is the **dependent variable** and time *t* is the **independent variable**. For this mathematical model there are 3 unknown parameters (*amplitude, decay_constant* and *background*). Possible sources of the background "noise" are cosmic radiation, noise in the instrumentation and sometimes a second much longer lived radioisotope within the source. The analysis will yield values for all three parameters but only the value of *decay_constant* is of interest. The half-life is determined from the resulting value of the decay constant:

$$e^{-decay\_constant \cdot half\_life} = 1/2$$

$$half\_life = \frac{0.69315}{decay\_constant} \qquad (1.1.2)$$

The number 0.69315 is the natural logarithm of 2. This mathematical model is based upon the physical phenomenon being observed: the number of counts recorded per unit time from the radioactive isotope decreases exponentially to the point where all that is observable is the background noise.

There are alternative methods for conducting and analyzing this experiment. For example, the value of **background** could be measured in a separate experiment. One could then subtract this value from the observed values of **Counts** and then use a mathematical model with only two unknown parameters (**amplitude** and **decay_constant**):

$$Counts - background = amplitude \cdot e^{-decay\_constant \cdot t} \qquad (1.1.3)$$

The selection of a mathematical model for a particular experiment might be trivial or it might be the main thrust of the work. Indeed, the purpose of many experiments is to either prove or disprove a particular mathematical model. If, for example, a mathematical model is shown to agree with experimental results, it can then be used to make predictions of the dependent variable for other values of the independent variables.

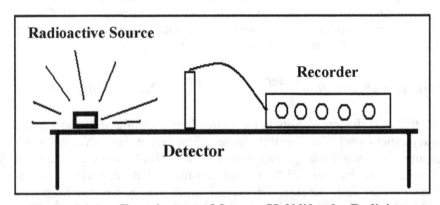

Figure 1.1.1   Experiment to Measure Half-life of a Radioisotope

Another important aspect of experimental work relates to the determination of the unknown parameters. Besides evaluation of these parameters by experiment, there might be an alternative calculation of the parameters based upon theoretical considerations. The purpose of the experiments for such cases is to confirm the theoretical results. Indeed, experiments go hand-in-hand with theory to improve our knowledge of the world around us.

Equations (1.1.1) and (1.1.3) are examples of mathematical models with only one independent variable (i.e., time t) and only one dependent variable (i.e., *Counts*). Often the mathematical model requires several independent variables and sometimes even several dependent variables. For example, consider classical chemical engineering experiments in which reaction rates are measured as functions of both pressure and temperature:

$$reaction\_rate = f(pressure, temperature) \tag{1.1.4}$$

The actual form of the function f is dependent upon the type of reaction being studied.

The following example relates to an experiment that requires two dependent variables. This experiment is a variation of the experiment illustrated in Figure 1.1.1. Some radioactive isotopes decay into a second radioisotope. The decays from both isotopes give off signals of different energies and appropriate instrumentation can differentiate between the two different signals. We can thus measure count rates from each isotope simultaneously. If we call them $c1$ and $c2$, assuming background radiation is negligible, the appropriate mathematical model would be:

$$c1 = a1 \cdot e^{-d1 \cdot t} \tag{1.1.5}$$

$$c2 = a2 \cdot e^{-d2 \cdot t} + a1 \frac{d2}{d2 - d1} \left(e^{-d1 \cdot t} - e^{-d2 \cdot t} \right) \tag{1.1.6}$$

This model contains four unknown parameters: the two amplitudes ($a1$ and $a2$) and the two decay constants ($d1$ and $d2$). The two dependent variables are $c1$ and $c2$, and the single independent variable is time t. The time dependence of $c1$ and $c2$ are shown in Figure 1.1.2 for one set of the parameters.

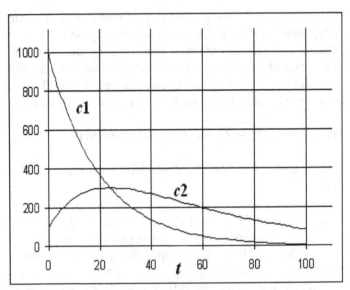

Figure 1.1.2   Counts versus Time for Equations 1.1.5 and 1.1.6
a1=1000, a2=100, d1=0.05, d2=0.025

The purpose of conducting experiments is not necessarily to prove or dis-prove a mathematical model or to determine parameters of a model. For some experiments the only purpose is to extract an equation from the data that can be used to predict values of the dependent variable (or variables) as a function of the independent variable (or variables). For such experi-ments the data is analyzed using different proposed equations (i.e., mathe-matical models) and the results are compared in order to select a "best" model.

We see that there are different reasons for performing quantitative experi-ments but what is common to all these experiments is the task of data analysis. In fact, there is no need to differentiate between physical ex-periments and experiments based upon computer generated data. Once data has been obtained, regardless of its origin, the task of data analysis commences. Whether or not the method of least squares is applicable de-pends upon the applicability of some basic assumptions. A discussion of the conditions allowing least squares analysis is included in Section 1.5: **Basic Assumptions.**

1.2 Dealing with Uncertainty

The estimation of uncertainty is an integral part of data analysis. It is not enough to just measure something. We always need an estimate of the accuracy of our measurements. For example, when we get on a scale in the morning, we know that the uncertainty is plus or minus a few hundred grams and this is considered acceptable. If, however, our scale were only accurate to plus or minus 10 kilograms this would be unacceptable. For other measurements of weight, an accuracy of a few hundred grams would be totally unacceptable. For example, if we wanted to purchase a gold bar, our accuracy requirements for the weight of the gold bar would be much more stringent. When performing quantitative experiments, we must take into consideration uncertainty in the input data. Also, the output of our analysis must include estimates of the uncertainty of the results. One of the most compelling reasons for using least squares analysis of data is that uncertainty estimates are obtained quite naturally as a part of the analysis. For almost all applications the standard deviation (σ) is the accepted measure of uncertainty. Let us say we need an estimate of the uncertainty associated with the measurement of the weight of gold bars. One method for obtaining such an estimate is to repeat the measurement n times and record the weights w_i , i = 1 to n. The estimate of σ (the estimated standard deviation of the weight measurement) is computed as follows:

$$\sigma^2 = \frac{1}{n-1} \sum_{i=1}^{i=n} (w_i - w_{avg})^2 \tag{1.2.1}$$

In this equation w_{avg} is the average value of the n measurements of w. The need for n-1 in the denominator of this equation is best explained by considering the case in which only one measurement of w is made (i.e., $n = 1$). For this case we have no information regarding the "spread" in the measured values of w.

Fortunately, for most measurements we don't have to estimate σ by repeating the measurement many times. Often the instrument used to perform the measurement is provided with some estimation of the accuracy of the measurements. Typically the estimation of σ is provided as a fixed percentage (e.g., $\sigma = 1\%$) or a fixed value (e.g., $\sigma = 0.5$ grams). Sometimes the accuracy is dependent upon the value of the quantity being measured in a more complex manner than just a fixed percentage or a constant value. For such cases the provider of the measurement instrument might supply this information in a graphical format or perhaps as an equation. For cases in which the data is calculated rather than measured, the calculation is incomplete unless it is accompanied by some estimate of uncertainty.

Once we have an estimation of σ, how do we interpret it? In addition to σ, we have a result either from measurements or from a calculation. Let us define the result as x and the true (but unknown value) of what we are trying to measure or compute as μ. Typically we assume that our best estimate of this true value of μ is x and that μ is located within a region around x. The size of the region is characterized by σ. A typical assumption is that the probability of μ being greater or less than x is the same. In other words, our measurement or calculation includes a random error characterized by σ. Unfortunately this assumption is not always valid!

Sometimes our measurements or calculations are corrupted by **systematic errors**. Systematic errors are errors that cause us to either systematically under-estimate or over-estimate our measurements or computations. One source of systematic errors is an unsuccessful calibration of a measuring instrument. Another source is failure to take into consideration external factors that might affect the measurement or calculation (e.g., temperature effects). Data analysis of quantitative experiments is based upon the assumption that the measured or calculated independent and dependent variables are not subject to systematic errors. If this assumption is not true, then errors are introduced into the results that do not show up in the computed values of the σ's. One can modify the least squares analysis to study the sensitivity of the results to systematic errors but whether or not systematic errors exist is a fundamental issue in any work of an experimental nature.

1.3 Statistical Distributions

In nature most quantities that are observed are subject to a statistical distribution. The distribution is often inherent in the quantity being observed but might also be the result of errors introduced in the method of observation. An example of an inherent distribution can be seen in a study in which the percentage of smokers is to be determined. Let us say that one thousand people above the age of 18 are tested to see if they are smokers. The percentage is determined from the number of positive responses. It is obvious that if 1000 different people are tested the result will be different. If many groups of 1000 were tested we would be in a position to say something about the distribution of this percentage. But do we really need to test many groups? Knowledge of statistics can help us estimate the standard deviation of the distribution by just considering the first group!

As an example of a distribution caused by a measuring instrument, consider the measurement of temperature using a thermometer. Uncertainty can be introduced in several ways:

1) The persons observing the result of the thermometer can introduce uncertainty. If, for example, a nurse observes a temperature of a patient as 37.4°C, a second nurse might record the same measurement as 37.5°C. (Modern thermometers with digital outputs can eliminate this source of uncertainty.)

2) If two measurements are made but the time taken to allow the temperature to reach equilibrium is different, the results might be different. (Taking care that sufficient time is allotted for the measurement can eliminate this source of uncertainty.)

3) If two different thermometers are used, the instruments themselves might be the source of a difference in the results. This source of uncertainty is inherent in the quality of the thermometers. Clearly, the greater the accuracy, the higher is the quality of the instrument and usually, the greater the cost. It is far more expensive to measure a temperature to 0.001°C than 0.1°C!

We use the symbol Φ to denote a distribution. Thus $\Phi(x)$ is the distribution of some quantity x. If x is a discrete variable then the definition of $\Phi(x)$ is:

$$\sum_{xmin}^{xmax} \Phi(x) = 1 \qquad\qquad (1.3.1)$$

If x is a continuous variable:

$$\int_{xmin}^{xmax} \Phi(x)dx = 1 \qquad\qquad (1.3.2)$$

Two important characteristics of all distributions are the mean μ and the variance σ^2. The standard deviation σ is the square root of the variance. For discrete distributions they are defined as follows:

$$\mu = \sum_{xmin}^{xmax} x\Phi(x) \tag{1.3.3}$$

$$\sigma^2 = \sum_{xmin}^{xmax} (x-\mu)^2\Phi(x) \tag{1.3.4}$$

For continuous distributions:

$$\mu = \int_{xmin}^{xmax} x\Phi(x)dx \tag{1.3.5}$$

$$\sigma^2 = \int_{xmin}^{xmax} (x-\mu)^2\Phi(x)\,dx \tag{1.3.6}$$

The normal distribution

When x is a continuous variable the normal distribution is often applicable. The normal distribution assumes that the range of x is from $-\infty$ to ∞ and that the distribution is symmetric about the mean value μ. These assumptions are often reasonable even for distributions of discrete variables, and thus the normal distribution can be used for some distributions of discrete variables. The equation for a normal distribution is:

$$\Phi(x) = \frac{1}{\sigma(2\pi)^{1/2}} exp(-\frac{(x-\mu)^2}{2\sigma^2}) \tag{1.3.7}$$

The normal distribution is shown in Figure 1.3.1 for various values of the standard deviation σ. We often use the term **standard normal distribution** to characterize one particular distribution: a normal distribution with mean $\mu = 0$ and standard deviation $\sigma = 1$. The symbol u is usually used to denote this distribution. Any normal distribution can be transformed into a standard normal distribution by subtracting μ from the values of x and then dividing this difference by σ.

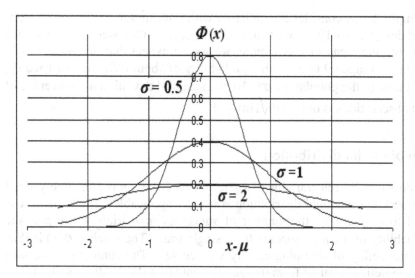

Figure 1.3.1 $\Phi(x)$ vs $x-\mu$ for Normal Distribution (σ=0.5, 1 and 2).

We can define the **effective range** of the distribution as the range in which a specified percentage of the data can be expected to fall. If we specify the effective range of the distribution as the range between $\mu \pm \sigma$, then 68.3% of all measurements would fall within this range. Extending the range to $\mu \pm 2\sigma$, 95.4% would fall within this range and 99.7% would fall within the range $\mu \pm 3\sigma$. The true range of any normal distribution is always -∞ to ∞. Values of the percentage that fall within 0 to u (i.e., $(x-\mu)/\sigma$) are included in tables in many sources [e.g., AB64, FR92]. The standard normal table is also available online [ST03]. Approximate equations corresponding to a given value of probability are also available (e.g., See Appendix B).

The normal distribution is not applicable for all distributions of continuous variables. In particular, if the variable x can only assume positive values and if the mean of the distribution μ is close to zero, then the normal distribution might lead to erroneous conclusions. If however, the value of μ is large (i.e., $\mu/\sigma \gg 1$) then the normal distribution is usually a good approximation even if negative values of x are impossible.

We are often interested in understanding how the mean of a sample of n values of x (i.e., x_{avg}) is distributed. It can be shown that the standard deviation of the value of x_{avg} has a standard deviation of $\sigma/\sqrt{n}$. Thus the quantity $(x_{avg}-\mu) / (\sigma/\sqrt{n})$ follows the standard normal distribution u. For example, let us consider a population with a mean value of 50 and a standard deviation of 10. If we take a sample of $n = 100$ observations and then compute the mean of this sample, we would expect that this mean would fall in the range 49 to 51 with a probability of about 68%. In other words, even though the population σ is 10, the standard deviation of an average of 100 observations is only $10/\sqrt{100} = 1$.

The binomial distribution

When x is a discrete variable of values 0 to n (where n is a relatively small number), the binomial distribution is usually applicable. The variable x is used to characterize the number of **successes** in n trials where p is the probability of a single success for a single trial. The symbol $\Phi(x)$ is thus the probability of obtaining exactly x successes. The number of successes can theoretically range from 0 to n. The equation for the distribution is:

$$\Phi(x) = \frac{n!}{x!(n-x)!} p^x (1-p)^{n-x} \qquad (1.3.8)$$

As an example, consider the following problem: what is the probability of drawing the Ace of Spades from a deck of cards if the total number of trials is 3. After each trial the card drawn is reinserted into the deck and the deck is shuffled. For this problem the possible values of x are 0, 1, 2 and 3. The value of p is 1/52 as there are 52 different cards in a deck: the Ace of Spades and 51 other cards. The probability of not drawing the Ace of Spades in any of the 3 trials is:

$$\Phi(0) = \frac{3!}{0!(3)!} p^0 (1-p)^3 = (51/52)^3 = 0.9434$$

The probability of drawing the Ace of Spades once is:

$$\Phi(1) = \frac{3!}{1!(2)!} p^1 (1-p)^2 = \frac{6}{2}(1/52)^1 (51/52)^2 = 0.0555$$

The probability of drawing the Ace of Spades twice is:

$$\Phi(2) = \frac{3!}{2!(1)!} p^2 (1-p)^1 = \frac{6}{2} (\tfrac{1}{52})^2 (\tfrac{51}{52})^1 = 0.00109$$

The probability of drawing the Ace of Spades all three times is:

$$\Phi(3) = \frac{3!}{3!(0)!} p^3 (1-p)^0 = (\tfrac{1}{52})^3 = 0.000007$$

The sum of all 4 of these probable outcomes is one. The probability of drawing the Ace of Spades at least once is 1 - 0.9434 = 0.0566.

The mean value μ and standard deviation σ of the binomial distribution can be computed from the values of n and p:

$$\mu = np \tag{1.3.9}$$

$$\sigma = (np(1-p))^{1/2} \tag{1.3.10}$$

Equation 1.3.9 is quite obvious. If, for example, we flip a coin 100 times, what is the average value of the number of heads we would observe? For this problem, $p = \tfrac{1}{2}$, so we would expect to see on average 100 * 1/2 = 50 heads. The equation for the standard deviation is not obvious, however the proof of this equation can be found in many elementary textbooks on statistics. For this example we compute σ as $(100*1/2*1/2)^{1/2} = 5$. Using the fact that the binomial distribution approaches a normal distribution for values of $\mu \gg 1$, we can estimate that if the experiment is repeated many times, the numbers of heads observed will fall within the range 45 to 55 about 68% of the time.

The Poisson distribution

The binomial distribution (i.e., Equation 1.3.8) becomes unwieldy for large values of n. The Poisson distribution is used for a discrete variable x that can vary from 0 to ∞. If we assume that we know the mean value μ of the distribution, then $\Phi(x)$ is computed as:

$$\Phi(x) = \frac{e^{-\mu} \mu^x}{x!} \tag{1.3.11}$$

It can be shown that the standard deviation σ of the Poisson distribution is:

$$\sigma = \mu^{1/2} \tag{1.3.12}$$

If μ is a large value, the normal distribution is an excellent approximation of a Poisson distribution.

As an example of a Poisson distribution, consider the observation of a rare genetic problem. Let us assume that the problem is observed on average 2.3 times per 10000 people. For practical purposes n is close to ∞ so we can assume that the Poisson distribution is applicable. We can compute the probability of observing x people with the genetic problem out of a sample population of 10000 people. The probability of observing no one with the problem is:

$$\Phi(0) = e^{-2.3}\, 2.3^0 / 0! = e^{-2.3} = 0.1003$$

The probability of observing one person with the problem is:

$$\Phi(1) = e^{-2.3}\, 2.3^1 / 1! = 2.3e^{-2.3} = 0.2306$$

The probability of observing two people with the problem is:

$$\Phi(2) = e^{-2.3}\, 2.3^2 / 2! = 2.3^2\, e^{-2.3} / 2 = 0.2652$$

The probability of observing three people with the problem is:

$$\Phi(3) = e^{-2.3}\, 2.3^3 / 3! = 2.3^3\, e^{-2.3} / 6 = 0.2136$$

From this point on, the probability $\Phi(x)$ decreases more and more rapidly and for all intents and purposes approaches zero for large values of x.

Another application of Poisson statistics is for counting experiments in which the number of counts is large. For example, consider observation of a radioisotope by an instrument that counts the number of signals emanating from the radioactive source per unit of time. Let us say that 10000 counts are observed. Our first assumption is that 10000 is our best estimate of the mean μ of the distribution. From equation 1.3.12 we can then estimate the standard deviation σ of the distribution as $10000^{1/2} = 100$. In other words, in a counting experiment in which 10000 counts are observed, the accuracy of this observed count rate is approximately 1% (i.e., 100/10000 = 0.01). To achieve an accuracy of 0.5% we can compute the required number of counts:

$$0.005 = \sigma / \mu = \mu^{1.2} / \mu = \mu^{-1/2}$$

Solving this equation we get a value of $\mu = 40000$. In other words to double our accuracy (i.e., halve the value of σ) we must increase the observed number of counts by a factor of 4.

The χ^2 distribution

The χ^2 (chi-squared) distribution is defined using a variable u that is normally distributed with a mean of 0 and a standard deviation of 1. This u distribution is called the standard normal distribution. The variable $\chi^2(k)$ is called the χ^2 value with k degrees of freedom and is defined as follows:

$$\chi^2(k) = \sum_{i=1}^{i=k} u_i^2 \tag{1.3.13}$$

In other words, if k samples are extracted from a standard normal distribution, the value of $\chi^2(k)$ is the sum of the squares of the u values. The distribution of these values of $\chi^2(k)$ is a complicated function:

$$\Phi(\chi^2(k)) = \frac{(\chi^2)^{k/2-1}}{2^{k/2}\Gamma(k/2)} \exp(-\chi^2/2) \tag{1.3.14}$$

In this equation Γ is called the gamma function and is defined as follows:

$$\Gamma(k/2) = (k/2-1)(k/2-2)...3*2*1 \; \textit{for k even}$$
$$\Gamma(k/2) = (k/2-1)(k/2-2)...3/2*1/2*\pi^{1/2} \; \textit{for k odd} \tag{1.3.15}$$

Equation 1.3.14 is complicated and rarely used. Of much greater interest is determination of a range of values from this distribution. What we are more interested in knowing is the probability of observing a value of χ^2 from 0 to some specified value. This probability can be computed from the following equation [AB64]:

$$P(\chi^2/k) = \frac{1}{2^{k/2}\,\Gamma(k/2)} \int_0^{\chi^2} t^{k/2-1} e^{-t/2}\,dt \qquad (1.3.16)$$

For small values of k (typically up to $k=30$) values of χ^2 are presented in a tabular format [e.g., AB64, FR92, ST03] but for larger values of k, approximate values can be computed (using the normal distribution approximation described below). The tables are usually presented in an inverse format (i.e., for a given value of k, the values of χ^2 corresponding to various probability levels are tabulated). As an example of the use of this distribution, let us consider an experiment in which we are testing a process to check if something has changed. Some variable x characterizes the process. We know from experience that the mean of the distribution of x is μ and the standard deviation is σ. The experiment consists of measuring 10 values of x. An initial check of the computed average value for the 10 values of x is seen to be close to the historical value of μ but can we make a statement regarding the variance in the data? We would expect that the following variable would be distributed as a standard normal distribution ($\mu=0$, $\sigma=1$):

$$u = \frac{(x-\mu)}{\sigma} \qquad (1.3.17)$$

Using Equation 1.3.17, 1.3.13 and the 10 values of x we can compute a value for χ^2. Let us say that the value obtained is 27.2. The question that we would like to answer is what is the probability of obtaining this value or a greater value by chance? From [ST03] it can be seen that for $k = 10$, there is a probability of 0.5% that the value of χ^2 will exceed 25.188. (Note that the value of k used was 10 and not 9 because the historical value of μ was used in Equation 1.3.17 and not the mean value of the 10 observations.) The value observed (i.e., 27.2) is thus on the high end of what we might expect by chance and therefore some problem might have arisen regarding the process under observation.

Two very useful properties of the χ^2 distribution are the mean and standard deviation of the distribution. For k degrees of freedom, the mean is k and the standard deviation is $\sqrt{2k}$. For large values of k, we can use the fact that this distribution approaches a normal distribution and thus we can easily compute ranges. For example, if $k = 100$, what is the value of χ^2 for which only 1% of all samples would exceed it by chance? For a standard normal distribution, the 1% limit is 2.326. The value for the χ^2 distribution would thus be $\mu + 2.326*\sigma = k + 2.326*(2k)^{1/2} = 100 + 31.2 = 131.2$.

An important use for the χ^2 distribution is analysis of variance. The **variance** is defined as the standard deviation squared. We can get an **unbiased estimate** of the variance of a variable x by using n observations of the variable. Calling this unbiased estimate as s^2, we compute it as follows:

$$s^2 = \frac{1}{n-1}\sum_{i=1}^{n}(x - x_{avg})^2 \qquad (1.3.18)$$

The quantity $(n-1)s^2/\sigma^2$ is distributed as χ^2 with n-1 degrees of freedom. This fact is fundamental for least squares analysis.

The t distribution

The t distribution (sometimes called the student-t distribution) is used for samples in which the standard deviation is not known. Using n observations of a variable x, the mean value x_{avg} and the unbiased estimate s of the standard deviation can be computed. The variable t is defined as:

$$t = (x_{avg} - \mu)/(s/\sqrt{n}) \qquad (1.3.19)$$

The t distribution was derived to explain how this quantity is distributed. In our discussion of the normal distribution, it was noted that the quantity $(x_{avg}-\mu)/(\sigma/\sqrt{n})$ follows the standard normal distribution u. When σ of the distribution is not known, the best that we can do is use s instead. For large values of n the value of s approaches the true value of σ of the distribution and thus t approaches a standard normal distribution. The mathematical form for the t distribution is based upon the observation that Equation 1.3.19 can be rewritten as:

$$t = \frac{(x_{avg} - \mu)}{(\sigma / \sqrt{n})} (\sigma/s) \qquad (1.3.20)$$

The term σ/s is distributed as $((n-1)/\chi^2)^{1/2}$ where χ^2 has n-1 degrees of freedom. Thus the mathematical form of the t distribution is derived from the product of the standard normal distribution and $((n-1)/\chi^2(n-1))^{1/2}$. Values of t for various percentage levels for n-1 up to 30 are included in tables in many sources [e.g., AB64, FR92]. The t table is also available online [ST03]. For values of $n > 30$, the t distribution is very close to the standard normal distribution.

For small values of n the use of the t distribution instead of the standard normal distribution is necessary to get realistic estimates of ranges. For example, consider the case of 4 observations of x in which x_{avg} and s of the measurements are 50 and 10. The value of $s/\sqrt{n}$ is 5. The value of t for $n - 1 = 3$ degrees of freedom and 1% is 4.541. We can use these numbers to determine a range for the true (but unknown value) of μ:

$$27.30 = 50 - 4.541 * 5 <= \mu <= 50 + 4.541 * 5 = 77.71$$

In other words, the probability of μ being below 27.30 is 1%, above 77.71 is 1% and within this range is 98%. Note that the value of 4.541 is considerably larger than the equivalent value of 2.326 for the standard normal distribution. It should be noted, however, that the t distribution approaches the standard normal rather rapidly. For example, the 1% limit is 2.764 for 10 degrees of freedom and 2.485 for 25 degrees of freedom. These values are only 19% and 7% above the standard normal 1% limit of 2.326.

The F distribution

The F distribution plays an important role in data analysis. This distribution was named to honor R.A. Fisher, one of the great statisticians of the 20th century. The F distribution is defined as the ratio of two χ^2 distributions divided by their degrees of freedom:

$$F = \frac{\chi^2(k_1)/k_1}{\chi^2(k_2)/k_2} \qquad (1.3.21)$$

The resulting distribution is complicated but tables of values of F for various percentage levels and degrees of freedom are available in many sources (e.g., [AB64, FR92]). Tables are also available online [ST03]. Simple equations for the mean and standard deviation of the F distribution are as follows:

$$\mu = \frac{k_2}{k_2 - 2} \quad \text{for } k_2 > 2 \tag{1.3.22}$$

$$\sigma^2 = \frac{2k_2^2(k_2 + k_1 - 2)}{k_1(k_2 - 2)^2(k_2 - 4)} \quad \text{for } k_2 > 4 \tag{1.3.23}$$

From these equations we see that for large values of k_2 μ approaches 1 and σ^2 approaches $2(1/k_1 + 1/k_2)$. If k_1 is also large, we see that σ^2 approaches zero. Thus if both k_1 and k_2 are large, we would expect the value of F to be very close to one.

1.4 Parametric Models

Quantitative experiments are usually based upon parametric models. In this discussion we define **parametric models** as models utilizing a mathematical equation that describes the phenomenon under observation. The model equation (or equations) contains unknown parameters and the purpose of the experiment is often to determine the parameters including some indication regarding the accuracy of these parameters. There are many situations in which the values of the individual parameters are of no interest. All that is important for these cases is that the parametric model can be used to predict values of the dependent variable (or variables) for other combinations of the independent variables. In addition, we are also interested in some measure of the accuracy of the predictions.

We need to use mathematical terminology to define parametric models. Let us use the term y to denote the dependent variable (or variables). Usually y is a scalar, but when there is more than one dependent variable, y can denote a vector. The parametric model is the mathematical equation that defines the relationship between the dependent and independent variables. For the case of a single dependent and a single independent variable we can denote the model as:

$$y = f(x; a_1, a_2 .., a_p) \tag{1.4.1}$$

The a_k's are the p unknown parameters of the model. The function f is based on either theoretical considerations or perhaps it is based on the behavior observed from the measured values of y and x.

When there is more than one independent variable, we can use the following to denote the model:

$$y = f(x_1, x_2 .., x_m; a_1, a_2 .., a_p) \tag{1.4.2}$$

The x_j's are the m independent variables. If there is more than one dependent variable, we require a separate function for each element of the y vector:

$$y_l = f_l(x_1, x_2 .., x_m; a_1, a_2 .., a_p) \quad l = 1 \text{ to } d \tag{1.4.3}$$

For cases of this type, y is a d dimensional vector and the subscript l refers to the lth term of the y vector. It should be noted that some or all of the x_j's and the a_k's may be included in each of the d equations. The notation for the ith data point for this l^{th} term would be:

$$y_{l_i} = f_l(x_{1_i}, x_{2_i} .., x_{m_i}; a_1, a_2 .., a_p)$$

Equations 1.1.5 and 1.1.6 illustrate an example of an experiment in which there are two dependent variables ($c1$ and $c2$), four unknown parameters ($a1, a2, d1$ and $d2$) and a single independent variable time t.

A model is recursive if the functions defining the dependent variables y_i are interdependent. The form for the elements of recursive models is as follows:

$$y_l = f_l(x_1, x_2 .., x_m; y_1, y_2 .., y_d; a_1, a_2 .., a_p) \tag{1.4.4}$$

As an example of a recursive model consider the following:

$$y_1 = a_1 x \sqrt{y_2} + a_2 \tag{1.4.5}$$

$$y_2 = a_3 x \sqrt{y_1} + a_4 \tag{1.4.6}$$

Both of these equations are recursive: there is one independent variable x, four unknown parameters (a_1 to a_4) and two dependent variables (y_1 and y_2). We see that y_1 is dependent upon y_2 and y_2 is dependent upon y_1.

Once a parametric model has been proposed and data is available, the task of data analysis must be performed. There are several possible objectives of interest to the analyst:

1) Compute the values of the p unknown parameters $a_1, a_2, ..., a_p$
2) Compute estimates of the standard deviations of the p unknown parameters.
3) Use the p unknown parameters to compute values of y for desired combinations of the independent variables $x_1, x_2, ..., x_m$
4) Compute estimates of the standard deviations σ_f for the values of $y = f(x)$ computed in 3.

It should be mentioned that the theoretically best solution to all of these objectives is achieved by applying the **method of maximum likelihood**. This method was proposed as a general method of estimation by the re-nowned statistician R. A. Fisher in the early part of the 20th century [e.g., FR92]. The method can be applied when the uncertainties associated with the observed or calculated data exhibit any type of distribution. However, when these uncertainties are normally distributed or when the normal dis-tribution is approximately correct, the method of maximum likelihood re-duces to the **method of least squares** [WO67, HA01]. A detailed proof of this statement is included in a book written by Merriman over 100 years ago [ME77]. Fortunately, the assumption of normally distributed random errors is reasonable for most situations and thus the method of least squares is applicable for analysis of most quantitative experiments.

1.5 Basic Assumptions

The method of least squares can be applied to a wide variety of analyses of experimental data. The common denominator for this broad class of prob-lems is the applicability of several basic assumptions. Before discussing these assumptions let us consider the measurement of a dependent variable Y_i. For the sake of simplicity, let us assume that the model describing the behavior of this dependent variable includes only a single independent variable. Using Equation 1.4.1 as the model that describes the relationship between x and y then y_i is the computed value of y at x_i. We define the difference between the measured and computed values as the residual R_i:

$$Y_i = y_i + R_i = f(x_i; a_1, a_2, \ldots a_p) + R_i \qquad (1.5.1)$$

It should be understood that neither Y_i nor y_i are necessarily equal to the true value η_i. In fact there might not even be a single true value if the dependent variable can only be characterized by a distribution. However, for the sake of simplicity let us assume that for every value of x_i there is a unique true value (or a unique mean value) of the dependent variable that is η_i. The difference between Y_i and η_i is the error ε_i:

$$Y_i = \eta_i + \varepsilon_i \qquad (1.5.2)$$

The development of the method of least squares in this book is based upon the following assumptions:

1) If the measurement at x_i were to be repeated many times, then the values of the error ε_i would be normally distributed with an average value of zero. Alternatively, if the errors are not normally distributed, the approximation of a normal distribution is reasonable.

2) The errors are uncorrelated. This is particularly important for time-dependent problems and implies that if a value measured at time t_i includes an error ε_i and at time t_{i+k} includes an error ε_{i+k} these errors are not related.

3) The standard deviations σ_i of the errors can vary from point to point. This assumption implies that σ_i is not necessarily equal to σ_j.

The implication of the first assumption is that if the measurement of Y_i is repeated many times, the average value of Y_i would be the true (i.e., error-less) value η_i. Furthermore, if the model is a true representation of the connection between y and x and if we knew the true values of the unknown parameters the residuals R_i would equal the errors ε_i:

$$Y_i = \eta_i + \varepsilon_i = f(x_i; \alpha_1, \alpha_2, \ldots \alpha_p) + \varepsilon_i \qquad (1.5.3)$$

In this equation the true value of the a_k is represented as α_k. However, even if the measurements are perfect (i.e., $\varepsilon_i = 0$), if f does not truly describe the dependency of y upon x, then there will certainly be a difference between the measured and computed values of y.

The first assumption of normally distributed errors is usually reasonable. Even if the data is described by other distributions (e.g., the binomial or Poisson distributions), the normal distribution is often a reasonable approximation. But there are problems where an assumption of normality causes improper conclusions. For example, in risk analysis the probability of catastrophic events might be considerably greater than one might predict using a normal distribution. To site one specific area, earthquake predictions require analyses in which normal distributions cannot be assumed. Another area that is subject to similar problems is the modeling of insurance claims. Most of the data represents relatively small claims but there are usually a small fraction of claims that are much larger, negating the assumption of normality. In this book such problems are not considered.

One might ask when the second assumption (i.e., uncorrelated errors) is invalid? There are areas of science and engineering where this assumption is not really reasonable and therefore the method of least squares must be modified to take error correlation into consideration [DA95]. Davidian and Giltinan discuss problem in the biostatistics field in which repeated data measurements are taken. For example, in clinical trials, data might be taken for many different patients over a fixed time period. For such problems we can use the term Y_{ij} to represent the measurement at time t_i for patient j. Clearly it is reasonable to assume that ε_{ij} is correlated with the error at time t_{i+1} for the same patient. In this book, no attempt is made to treat such problems.

Many statistical textbooks include discussions of the method of least squares but use the assumption that all the σ_i's are equal. This assumption is really not necessary as the additional complexity of using varying σ_i's is minimal. Another simplifying assumption often used is that the models are linear with respect to the a_k's. This assumption allows a very simple mathematical solution but is too limiting for the analysis of many real-world experiments. This book treats the more general case in which the function f (or functions f_i) can be nonlinear.

1.6 Systematic Errors

I first became aware of systematic errors while doing my graduate research. My thesis was a study of the fast fission effect in heavy water nuclear reactors and I was reviewing previous measurements of this effect [WO62]. Experimental results from two of the national laboratories were curiously different. Based upon the values and quoted σ's, the numbers were many σ's apart. I discussed this with my thesis advisors and we agreed that one or both of the experiments was plagued by systematic errors that biased the results in a particular direction. We were proposing a new method which we felt was much less prone to systematic errors.

One of the basic assumptions mentioned in the previous section is that the errors in the data are random about the true values. In other words, if a measurement is repeated n times, the average value would approach the true value as n approaches infinity. However, what happens if this assumption is not valid? We call such errors **systematic errors** and they will of course cause errors in the results. Systematic errors can be introduced in a variety of ways. For example, if an experiment lasting several days is undertaken, the results might be temperature dependent. If there is a significant change in temperature this might result in serious errors in the results. A good experimentalist will consider what factors might affect the results of a proposed experiment and then take steps to either minimize these factors or take them into consideration as part of the proposed model.

We can make some statements about combining estimates of systematic errors. Let us assume that we have identified *nsys* sources of systematic errors and that we can estimate the maximum size of each of these error sources. Let us define ε_{jk} as the systematic error in the measurement of a_j caused by the kth source of systematic errors. The magnitude of the value of ε_j (the magnitude of the systematic error in the measurement of a_j caused by all sources) could range from zero to the sum of the absolute values of all the ε_{jk}'s. However, a more realistic estimate of ε_j is the following:

$$\varepsilon_j^2 = \sum_{k=1}^{k=nsys} \varepsilon_{jk}^2 \qquad (1.6.1)$$

In the following chapter the method of least squares is developed and estimates of the uncertainties in the results of the least squares analysis are included.   It should be remembered that the basic assumption of the method is that the data is not plagued by systematic errors. For example, let us say we use the method of least squares to determine the constants a_1 and a_2' of a straight line (i.e., $y = a_1 + a_2 x$) and let us say that the results indicate that the uncertainties σ_{a1} and σ_{a2} have been determined to 1% accuracy.   Let us also say that we estimate that ε_1 is approximately equal to $C_1 \sigma_{a1}$ and ε_2 is approximately equal to $C_2 \sigma_{a2}$, how do we report our results? If we assume independence of σ_{aj} and ε_j a more accurate estimate of the uncertainties associated with the results is:

$$\sigma_j^2 = \sigma_{aj}^2 + \varepsilon_j^2 = (1 + C_j^2)\sigma_{aj}^2 \tag{1.6.2}$$

Clearly, the experimentalist should make an effort to ensure that the ε_j's are small when compared to the σ_{aj}'s (i.e, make the C_j's as small as possible).

One might ask the question: how do I go about estimating the values of the ε_j's? Often it is possible to use the least squares software to help make these estimates. For example, if you suspect the maximum systematic error in the values of the x's is δ_x, then you could change all the x's by δ_x and repeat the least squares analysis. Comparing the results with the previous analysis reveals how the results are affected by the δ_x change. This is in fact a direct measurement of the ε's associated with this type of error. Note that δ_x is not the same as σ_x. The values of σ_x are random errors whereas the systematic error is a fixed error in all the values of x somewhere in the range $\pm\delta_x$.   For the straight-line fit, we can see from Figure 1.6.1 that the effect of a systematic error of magnitude δ_x in the values of x will cause a contribution to ε_1 equal to $-a_2\delta_x$ and will have no effect upon the value of a_2 (i.e., $\varepsilon_2 = 0$). Similarly, a systematic error of magnitude δ_y in the values of y will cause a contribution to ε_1 equal to δ_y and will have no effect upon the value of a_2 (i.e., $\varepsilon_2 = 0$). Assuming the these effects are independent, we can combine these two sources of systematic error to estimate a value for ε_1 :

$$\varepsilon_1^2 = a_2^2 \delta_x^2 + \delta_y^2 \tag{1.6.3}$$

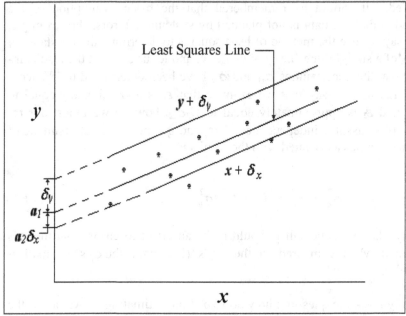

Figure 1.6.1   Effect of systematic error in x (δ_x) and in y (δ_y)

1.7 Nonparametric Models

There are situations in which it is quite useless to attempt to describe the phenomenon under observation by a single equation. For example, consider a dependent variable that is the future percentage return on stocks traded on the NYSE (New York Stock Exchange). One might be interested in trying to find a relationship between the future returns and several indicators that can be computed using currently available data. For this problem there is no underlying theory upon which a parametric model can be constructed. A typical approach to this problem is to allow the historic data to define a surface and then use some sort of smoothing technique to make future predictions regarding the dependent variable. The data plus the algorithm used to make the predictions are the major elements in what we define as a **nonparametric model.**

Nonparametric methods of data modeling predate the modern computer era [WO00]. In the 1920's two of the most well-known statisticians (Sir R. A. Fisher and E. S. Pearson) debated the value of such methods [HA90]. Fisher correctly pointed out that a parametric approach is inherently more efficient. Pearson was also correct in stating that if the true relationship between X and Y is unknown, then an erroneous specification in the function $f(X)$ introduces a model bias that might be disastrous.

Hardle includes a number of examples of successful nonparametric models [HA90]. The most impressive is the relationship between change in height (cm/year) and age of women (Figure 1.7.1). A previously undetected growth spurt at around age 8 was noted when the data was modeled using a nonparametric smoother [GA84]. To measure such an effect using parametric techniques, one would have to anticipate this result and include a suitable term in $f(X)$.

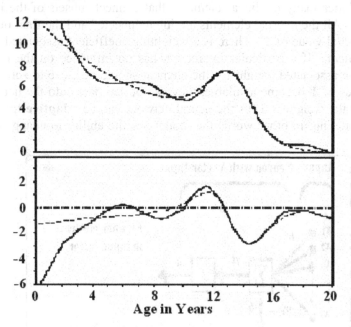

Figure 1.7.1   Human growth in women versus Age. The top graph is in cm/year. The bottom graph is acceleration in cm/year^2. The solid lines are from a model based upon nonparametric smoothing and the dashed lines are from a parametric fit [GA84, HA90].

Clearly, one can combine nonparametric and parametric modeling techniques. A possible strategy is to use nonparametric methods on an exploratory basis and then use the results to specify a parametric model. However, as the dimensionality of the model and the complexity of the surface increases, the hope of specifying a parametric model becomes more and more remote. An example of a problem area where parametric methods of modeling are not really feasible is the area of financial market modeling. As a result, there is considerable interest in applying nonpara-metric methods to the development of tools for making financial market predictions. A number of books devoted to this subject have been written in recent years (e.g., [AZ94, BA94, GA95, RE95, WO00]).

The emphasis on neural networks as a nonparametric modeling tool is par-ticularly attractive for time series modeling. The basic architecture of a single element (called a neuron) in a neural network is shown in Figure 1.7.2. The input vector X may include any number of variables. The net-work includes many nonlinear elements that connect subsets of the input variables. All the internal elements are interconnected and the final output is a predicted value of Y. There is a weighting coefficient associated with each element. If a particular interaction has no influence on the model output, the associated weight for the element should be close to zero. As new values of Y become available, they can be fed back into the network to update the weights. Thus the neural network can be **adaptive** for time series modeling: in other words the model has the ability to change over time.

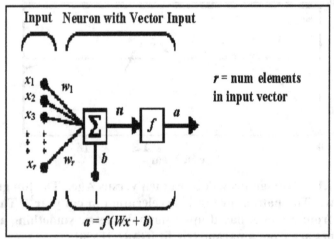

Figure 1.7.2   A typical element in a neural network. The Σ block sums the weighted inputs and the bias b. The f block is a nonlinear transfer function.

There is one major problem associated with neural network modeling: the amount of computer time required to generate a model. If the model is to be based upon a small number of predictor variables, then even if the number of data records is large, the required computer time is usually manageable. However, if one wishes to use tens or even hundreds of thousands of data records and hundreds of candidate predictors, the required computer time can be monumental. If one is to have any hope of success, techniques are required to preprocess the data in such a manner as to reduce the number of candidate predictors to a reasonable number. The definition of **reasonable** varies, of course, depending upon the available computing power. However, regardless of the hardware available, preprocessing strategies are essential to successfully apply neural nets to such problems (for example, financial market modeling). Use of kernel regression is an alternative modeling strategy that can be many orders of magnitude faster than the more compute intensive methods such as neural networks. It is certainly not as adaptive as neural networks but it can be used to very rapidly obtain the **information rich** subsets of the total candidate predictor space. These subspaces can, in turn be used as inputs to a neural network modeling program. The kernel regression method is based upon least squares and is discussed in detail in Chapter 7.

1.8 Statistical Learning

The term *statistical learning* is used to cover a broad class of methods and problems that have become feasible as the power of the computer has grown. An in-depth survey of this field is covered in a fairly recent book by Hastie, Tibshirani and Friedman entitled *The Elements of Statistical Learning: Data Mining, Inference and Prediction* [HA01]. Their book covers both supervised and unsupervised learning. The goal of supervised learning is to predict an output variable as a function of a number of input variables (or as they are sometimes called: indicators or predictors). In unsupervised learning there is no particular output variable and one is interested in finding associations and patterns among the variables. The cornerstone of statistical learning is to *learn from the data*. The analyst has access to data and his or her goal is to make sense out of the available information.

Supervised learning problems can be subdivided into *regression* and *classification* problems. The goal in regression problems is to develop quantitative predictions for the dependent variable. The goal in classification problems is to develop methods for predicting to which class a particular data point belongs. An example of a regression problem is the development of a model for predicting the unemployment rate as a function of economic indictors. An example of a classification problem is the development of a model for predicting whether or not a particular email message is a spam message or a real message. In this book, although classification problems are discussed (see Sections 2.8 and 7.8), the emphasis is on regression problems.

In this book we assume that the structure of the data is known (to some degree). Many problems in science and engineering fall within this category. Our typical starting point is a model that describes the relationship between the dependent and independent variables. The model includes some unknown parameters or parameters that we wish to determine to a greater accuracy than currently known. Sometimes the purpose of the experiment is to use the data as the basis for comparing different theoretical approaches to a particular problem. Sometimes the purpose of the experiment is to develop an equation that can be used for interpolation or extrapolation. The data might lead us to a modified form of the model, but the existence of a fair degree of structure is assumed.

In data mining applications, there are often a fairly large number of candidate predictors and the structure of the relationship between the dependent variable and the candidate predictors is not known or hardly known. The data miner searches for *information-rich* subsets of the candidate predictors that can be used for making predictions. One problem associated with such efforts is the *curse of dimensionality*, a concept first identified by Bellman in 1961 [BE61]. As the number of candidate predictors increases, the density of the data decreases exponentially. Stated in an alternative manner, if we wish to maintain the density of the data, for every added predictor, we must double the number of data points. Thus the number of available data points limits the number of candidate predictors that can be included in a model. As a result, data mining software typically includes methods for identifying the information-rich lower dimensional subsets of the total candidate predictor space. These subsets can be used individually for making predictions or can be combined to create a super-model if such a combination proves to be a better predictor than the individual subset models.

In this book, we assume that we know which independent variables must be included in the final model. For example, when studying chemical reaction rates, we consider the effects of temperature, pressure and time upon the concentrations of the chemical species undergoing the reaction. We know that the problem is dependent upon these variable and we also know the mathematical model relating the dependent variables with these independent variables. What we would like to know are some of the parameters that are included within the model. The approach to problems of this type is fundamentally different than the model searching techniques used in data mining applications.

Chapter 2 THE METHOD OF LEAST SQUARES

2.1 Introduction

The first published treatment of the method of least squares was included in an appendix to Adrien Marie Legendre's book *Nouvelles methods pour la determination des orbites des cometes*. The 9 page appendix was entitled *Sur la methode des moindres quarres*. The book and appendix was published in 1805 and included only 80 pages but gained a 55 page supplement in 1806 and a second 80 page supplement in 1820 [ST86]. It has been said that the method of least squares was to statistics what calculus had been to mathematics. The method became a standard tool in astronomy and geodesy throughout Europe within a decade of its publication. The method was also the cause of a dispute between two giants of the scientific world of the 19th century: Legendre and Gauss. Gauss in 1809 in his famous *Theoria Motus* claimed that he had been using the method since 1795. That book was first translated into English in 1857 under the authority of the United States Navy by the Nautical Almanac and Smithsonian Institute [GA57]. Another interesting aspect of the method is that it was rediscovered in a slightly different form by Sir Francis Galton. In 1885 Galton introduced the concept of regression in his work on heredity. But as Stigler says: "Is there more than one way a sum of squared deviations can be made small?" Even though the method of least squares was discovered about 200 years ago, it is still "the most widely used nontrivial technique of modern statistics" [ST86].

The least squares method is discussed in many books but the treatment is usually limited to linear least squares problems. In particular, the emphasis is often on fitting straight lines or polynomials to data. The multiple linear regression problem (described below) is also discussed extensively (e.g., [FR92, WA93]). Treatment of the general nonlinear least squares problem is included in a much smaller number of books. One of the earliest books on this subject was written by W. E. Deming and published in the pre-computer era in 1943 [DE43]. An early paper by R. Moore and R. Zeigler discussing one of the first general purpose computer programs for solving nonlinear least squares problems was published in 1960 [MO60]. The program described in the paper was developed at the Los Alamos Laboratories in New Mexico. Since then general least squares has been covered in varying degrees and with varying emphases by a number of authors (e.g., DR66, WO67, BA74, GA94, VE02).

For most quantitative experiments, the method of least squares is the "best" analytical technique for extracting information from a set of data. The method is best in the sense that the parameters determined by the least squares analysis are normally distributed about the true parameters with the least possible standard deviations. This statement is based upon the assumption that the uncertainties (i.e., errors) in the data are uncorrelated and normally distributed. For most quantitative experiments this is usually true or is a reasonable approximation. When the curve being fitted to the data is a straight line, the term **linear regression** is often used. For the more general case in which a plane based upon several independent variables is used instead of a simple straight line, the term **multiple linear regression** is often used [FR92, WA93]. Prior to the advent of the digital computer, curve fitting was usually limited to lines and planes. For the simplest problem (i.e., a straight line), the assumed relationship between the dependent variable y and the independent variable x is:

$$y = a_1 + a_2 x \tag{2.1.1}$$

For the case of more than one independent variable (multiple linear regression), the assumed relationship is:

$$y = a_1 x_1 + a_2 x_2 + \ldots + a_m x_m + a_{m+1} \tag{2.1.2}$$

For this more general case each data point includes $m+1$ values: $y_i, x_{1i}, x_{2i}, \ldots, x_{m_i}$.

The least squares solutions for problems in which Equations 2.1.1 and 2.1.2 are valid fall within the much broader class of **linear least squares** problems. In general, all linear least squares problem are based upon an equation of the following form:

$$y = f(\mathbf{X}) = \sum_{k=1}^{k=p} a_k g_k(\mathbf{X}) = \sum_{k=1}^{k=p} a_k g_k(x_1, x_2 \dots x_m) \qquad (2.1.3)$$

In other words, y is a function of $\mathbf{X}$ (a vector with m terms). Any equation in which the p unknown parameters (i.e., the a_k's) are coefficients of functions of only the independent variables (i.e., the m terms of the vector $\mathbf{X}$) can be treated as a linear problem. For example in the following equation, the values of a_1, a_2, and a_3 can be determined using linear least squares:

$$y = a_1 \sin(x_1^{3/2}) / \cosh(x-1) + a_2 (\cos(x_2^{5/2} - x^3))^{3/2}$$
$$+ a_3 / \ln(1/x + 1/x^2)$$

This equation is nonlinear with respect to x but the equation is linear with respect to the a_k's. In this example, the $\mathbf{X}$ vector contains only one term so we use the notation x rather than x_1. The following example is a linear equation in which $\mathbf{X}$ is a vector containing 2 terms:

$$y = a_1 \sin(x_1^{3/2}) / \cosh(x_2 - 1) + a_2 (\cos(x_1^{5/2} - x_2^3))^{3/2}$$
$$+ a_3 / \ln(1/x_1 + 1/x_1^2)$$

The following example is a nonlinear function:

$$y = a_1 \sin(x_1^{3/2}) / \cosh(x_2 - 1) + a_2 (\cos(x_1^{5/2} - x_2^3))^{3/2}$$
$$+ a_3 / \ln(1/x_1 + a_4/x_1^2)$$

The fact that a_4 is embedded within the last term makes this function incompatible with Equation 2.1.3 and therefore it is nonlinear with respect to the a_k's.

For both linear and nonlinear least squares, a set of p equations and p unknowns is developed. If Equation 2.1.3 is applicable then this set of equations is linear and can be solved directly. However, for nonlinear equations, the p equations require estimates of the a_k's and therefore iterations are required to achieve a solution. For each iteration, the a_k's are updated, the terms in the p equations are recomputed and the process is continued until some convergence criterion is achieved. Unfortunately, achieving convergence is not a simple matter for some nonlinear problems.

For some problems our only interest is to compute $y = f(\mathbf{X})$ and perhaps some measure of the uncertainty associated with these values (e.g., σ_f) for various values of $\mathbf{X}$. This is what is often called the **prediction** problem. We use measured or computed values of x and y to determine the parameters of the equation (i.e., the a_k's) and then apply the equation to calculate values of y for any value of x. For cases where there are several (let us say m) independent variables, the resulting equation allows us to predict y for any combination of $x_1, x_2, .., x_m$. The least squares formulation developed in this chapter also includes the methodology for prediction problems.

2.2 The Objective Function

The starting point for the method of least squares is the **objective function**. Minimization of this function yields the least squares solution. The simplest problems are those in which y (a scalar quantity) is related to an independent variable x (or variables x_j's) and it can be assumed that there is no (or negligible) errors in the independent variable (or variables). The objective function for these cases is:

$$S = \sum_{i=1}^{i=n} w_i R_i^2 = \sum_{i=1}^{i=n} w_i (Y_i - y_i)^2 = \sum_{i=1}^{i=n} w_i (Y_i - f(X_i))^2 \qquad (2.2.1)$$

In this equation n is the number of data points, Y_i is the ith input value of the dependent variable and y_i is the ith computed value of the dependent variable. The variable R_i is called the ith residual and is the difference between the input and computed values of y for the ith data point. The variable X_i (unitalicized) represents the independent variables and is either a scalar if there is only one independent variable or a vector if there is more than one independent variable. The function f is the equation used to express the relationship between X and y. The variable w_i is called the "weight" associated with the ith data point and is discussed in the next section. A schematic diagram of the variables for point i is shown in Figure 2.2.1. In this diagram there is only a single independent variable so the notation x is used instead of X. The variable E_i is the true but unknown error in the ith value of y. Note that neither the value of Y_i nor y_i is exactly equal to the unknown η_i (the true value of y) at this value of x_i. However, a fundamental assumption of the method of least squares is that if Y_i is determined many times, the average value would approach this true value.

The next level of complexity is when the uncertainties in the measured or calculated values of x are not negligible. The relevant schematic diagram is shown in Figure 2.2.2. For such cases the objective function must also include residuals in the x as well as the y direction:

$$S = \sum_{i=1}^{i=n} (w_{y_i} R_{y_i}^2 + w_{x_i} R_{x_i}^2)$$

$$= \sum_{i=1}^{i=n} (w_{y_i} (Y_i - y_i)^2 + w_{x_i} (X_i - x_i)^2)$$

(2.2.2)

In this equation, X_i (italicized) is the measured value of the ith independent variable and x_i is the computed value. Note that X_i is not the same as X_i in Equation 2.2.1. In that equation capital X (unitalicized) represents the vector of independent variables.

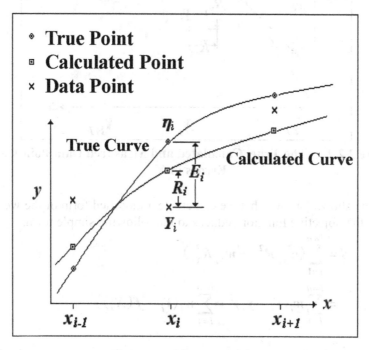

Figure 2.2.1   The True, Calculated and Measured Data Points with no Error in x

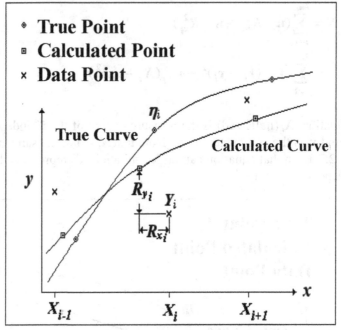

Figure 2.2.2   The True, Calculated and Measured Data Points with Errors in X

It can be shown [WO67] that we can create a modified form of the weights so that the objective function reduces to the following simple form:

$$S = \sum_{i=1}^{i=n} (w_{y_i} R_{y_i}^2 + w_{x_i} R_{x_i}^2)$$

$$= \sum_{i=1}^{i=n} w_i (Y_i - y_i)^2 = \sum_{i=1}^{i=n} w_i (Y_i - f(X_i))^2$$

(2.2.3)

In other words, Equation 2.2.1 is valid even if the uncertainties in the x variables are not negligible. All that is required is a modified form of the weighting function used to determine the values of w_i. Note that if there is more than one independent variable, an additional summation is required:

$$S = \sum_{i=1}^{i=n} \left(w_{y_i} R_{y_i}^2 + \sum_{j=1}^{j=m} w_{x_i} R_{x_{ji}}^2 \right) = \sum_{i=1}^{i=n} w_i (Y_i - f(X_i))^2$$

(2.2.4)

Note that in Eq 2.2.4 $\mathbf{X}_i$ is unitalicized because it represents the vector of the independent variables. The italicized X_i used in Eq 2.2.3 represents the scalar independent variable. If y is a vector quantity, then we must further modify the objective function by including a sum over all the y terms. Assuming that there are d terms in the y vector (i.e., y_i is a d dimensional vector), the objective function is:

$$S = \sum_{l=1}^{l=d} S_l = \sum_{l=1}^{d}\sum_{i=1}^{n} w_{l_i}(Y_{l_i} - y_{l_i})^2 = \sum_{l=1}^{d}\sum_{i=1}^{i=n} w_{l_i}(Y_{l_i} - f_l(\mathbf{X}_i))^2 \quad (2.2.5)$$

In a later section we discuss treatment of prior estimates of the unknown a_k parameters. To take these prior estimates into consideration we merely make an additional modification of the objective function. For example, assume that for each of the p unknown a_k's there is a prior estimate of the value. Let us use the notation b_k as the prior estimate of a_k and σ_{b_k} as the uncertainty associated with this prior estimate. In the statistical literature these prior estimates are sometimes called **Bayesian** estimators. (This terminology stems from the work of the Reverend Thomas Bayes who was a little known statistician born in 1701. Some of his papers eventually reached the Royal Society but made little impact until the great French mathematician Pierre Laplace discovered them there.) The modified form of Equation 2.2.1 is:

$$S = \sum_{i=1}^{i=n}\sum_{i=1}^{i=n} w_i(Y_i - f(\mathbf{X}_i))^2 + \sum_{k=1}^{k=p}(a_k - b_k)^2/\sigma_{b_k}^2 \quad (2.2.6)$$

If there is no Bayesian estimator for a particular a_k the value of σ_{b_k} is set to infinity.

Regardless of the choice of objective function and scheme used to determine the weights w_i, one must then determine the values of the p unknown parameters a_k that minimize S. To accomplish this task, the most common procedure is to differentiate S with respect to all the a_k's and the resulting expressions are set to zero. This yields p equations that can then be solved to determine the p unknown values of the a_k's. A detailed discussion of this process is included in Section 2.4. An alternative class of methods to find a "best" set of a_k's is to use an intelligent search within a limited range of the unknown parameter space. A number of such stochastic algorithms are discussed in the literature (e.g., TV04).

2.3 Data Weighting

In Section 2.2, we noted that regardless of the choice of the objective function, a weight w_i is specified for each point. The "weight" associated with a point is based upon the relative uncertainties associated with the different points. Clearly, we must place greater weight upon points that have smaller uncertainties, and less weight upon the points that have greater uncertainties. In other words the weight w_i must be related in some way to the uncertainties σ_{y_i} and σ_{x_i}.

The alternative to using w_i's associated with the σ's of the ith data point is to simply use **unit weighting** (i.e., $w_i=1$) for all points. This is a reasonable choice for w_i if the σ's for all points are approximately the same or if we have no idea regarding the values (actual or even relative) of σ for the different points. However, when the differences in the σ's are significant, then use of unit weighting can lead to poor results. This point is illustrated in Figure 2.3.1. In this example, we fit a straight line to a set of data. Note that the line obtained when all points are equally weighted is very different than the line obtained when the points are "weighted" properly. Also note how far the unit weighting line is from the first few points.

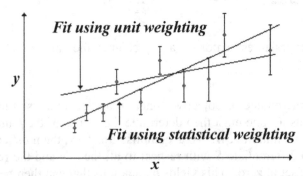

Figure 2.3.1   Two least squares lines with different weighting schemes

The question that must be answered is how do we relate w_i to the σ's associated with the ith data point? In Section 2.2 we noted that the objective function is of the form:

$$S = \sum_{i=1}^{i=n} S_i = \sum_{i=1}^{i=n} w_i R_i^2$$

$$= \sum_{i=1}^{i=n} w_i (Y_i - y_i)^2 = \sum_{i=1}^{i=n} w_i (Y_i - f(X_i))^2 \qquad (2.3.1)$$

We will see that the least squares solution is based upon adjusting the unknown values of the a_k's that are included in the function f such that S is minimized. **If the function f is representative of the data**, this minimization process yields values of S_i that tend to be distributed around an average value with some random error ε_i:

$$S_i = w_i R_i^2 = S_{\text{avg}} + \varepsilon_i \qquad (2.3.2)$$

For cases in which the uncertainties associated with the values of x_i are negligible our objective should be that the residuals R_i are proportional to the values of σ_{y_i}. If we define the relative error at point i (*Rel_Error$_i$*) as R_i/σ_{y_i}, our objective should be to have relative errors randomly distributed about 0. To accomplish this, we choose the following weighting function:

$$w_i = 1/\sigma_{y_i}^2 \qquad (2.3.3)$$

We call this type of weighting **statistical weighting** and we will see that it has many attractive properties. Substituting Equation 2.3.3 into 2.3.1 and 2.3.2 we obtain the following expression:

$$(R_i / \sigma_{y_i})^2 = (Rel\_Error_i)^2 = S_i = S_{\text{avg}} + \varepsilon_i \qquad (2.3.4)$$

We define **RMS(R)** as the "root mean square" error:

$$RMS(R) = \left(\sum_{i=1}^{i=n} R_i^2 \right)^{1/2} \qquad (2.3.5)$$

What we expect is that **RMS(R)** approaches zero as the noise component of the **y** values approaches zero. In reality, can we expect this from the least squares analysis? The answer to this question is yes but only if several conditions are met:

1) The function f is representative of the data. In other words, the data falls on the curve described by the function f with only a random "noise" component associated with each data point. For the case where the data is truly represented by the function f (i.e., there is no "noise" component to the data), then all the values of R_i will be zero and thus the values of S_i will be zero.

2) There are no erroneous data points or if data errors do exist, they are not significantly greater than the expected noise component. For example, if a measuring instrument should be accurate to 1%, then errors several times larger than 1% would be suspicious and perhaps problematical. If some points were in error far exceeding 1% then the results of the analysis will probably lead to significant errors in the final results. Clearly, we would hope that there are methods for detecting erroneous data points. This subject is discussed in detail in Chapter 3 in the section dealing with Goodness-of-Fit.

To illustrate the first point, consider the data shown in Table 2.3.1. This data was created using the following model:

$$Y = (7 - 6x + x^2) * (1 + 0.05u_i) \qquad (2.3.6)$$

Ten values of x were chosen as 1 thru 10 and the values of u_i were taken randomly from a u (i.e., standard normal) distribution. In other words, this model is a parabola with a noise component that is approximately 5% of the "true" value of y. It should be emphasized that we don't know and can't measure the values of u_i. All that we know is that the values of Y include about 5% random noise. Our weighting function for this case would be:

$$w_i = \frac{1}{\sigma_{y_i}^2} = \frac{1}{(0.05 * Y_i)^2}$$

The same data is shown in graphical form in Figure 2.3.2. Fitting a parabola to this data using the method of least squares we get the following curve:

$$y = f(x) = a_1 + a_2 x + a_3 x^2 = 7.086 - 6.026x + 0.999x^2$$

The values of the relative errors are seen in Table 2.3.1 to vary from -1.04 to 1.483 and seem to be distributed around zero as one would expect because the chosen model is representative of the data. In Table 2.3.2 we fit the data using a straight line and get the following curve:

$$y = f(x) = a_1 + a_2 x = 1.661 + 0.516x$$

The values of the relative errors are seen to be much greater (from -17.86 to 30.37) and they are clearly not distributed randomly about zero. This type of result is due to the choice of a function f that is clearly not representative of the data.

x	Y	σ_y	y	Rel_Error
1.0	2.047	0.102	2.060	-0.126
2.0	-0.966	0.048	-0.967	0.023
3.0	-1.923	0.096	-1.995	0.750
4.0	-1.064	0.053	-1.024	-0.749
5.0	2.048	0.102	1.946	0.998
6.0	6.573	0.329	6.915	-1.040
7.0	13.647	0.682	13.883	-0.345
8.0	24.679	1.234	22.850	1.483
9.0	34.108	1.705	33.816	0.171
10.0	44.969	2.248	46.780	-0.806

Table 2.3.1   Data generated using Equation 2.3.6 and fit using $y = a_1 + a_2 x + a_3 x^2$. Rel_Error is $(Y - y) / \sigma_y$.

x	Y	σ_y	y	Rel_Error
1.0	2.047	0.102	-1.062	30.37
2.0	-0.966	0.048	-0.746	-4.56
3.0	-1.923	0.096	-0.430	-15.54
4.0	-1.064	0.053	-0.114	-17.86
5.0	2.048	0.102	0.202	18.03
6.0	6.573	0.329	0.518	18.42
7.0	13.647	0.682	0.834	18.78
8.0	24.679	1.234	1.150	19.07
9.0	34.108	1.705	1.466	19.14
10.0	44.969	2.248	1.781	19.21

Table 2.3.2   Data generated using Equation 2.3.6 and fit using
$$y = a_1 + a_2x. \quad \textbf{\textit{Rel\_Error}} \text{ is } (Y - y) / \sigma_y.$$

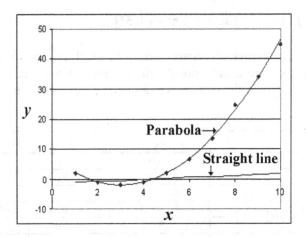

Figure 2.3.2   Parabola: Table 2.3.1,   Straight Line: Table 2.3.2

The effect of erroneous data points is discussed in the section on "outliers". (Section 3.6). This is a real concern of all experimentalists and care should be taken to search for outliers in any data set. The cause of outliers can be errors in data collections, errors in computer pre-processing of the data or even human error. However, one must not overlook the possibility that the outliers are real (i.e., not errors) and are perhaps due to an unexpected phe-nomenon. As an example of the effect of an erroneous data point, using the data from the previous example, the point Y_5 was changed from 2.048 to 20.48. This can happen, for example, when data is being collected by hand and the person recording the data misplaces the decimal point. The results of a least-squares analysis of this data are shown in Table 2.3.3.

Note the very large value of the relative error for point 5. In the discussion of Goodness-of-Fit (Section 3.2), statistical tests are discussed that allow us to measure the goodness of the fit without the need to examine all the relative errors.

x	Y	σ_y	y	Rel_Error
1.0	2.047	0.102	2.061	-0.135
2.0	-0.966	0.048	-0.972	0.118
3.0	-1.923	0.096	-1.993	0.724
4.0	-1.064	0.053	-1.002	-1.168
5.0	20.480	1.024	2.000	18.046
6.0	6.573	0.329	7.014	-1.343
7.0	13.647	0.682	14.040	-0.576
8.0	24.679	1.234	23.077	1.298
9.0	34.108	1.705	34.126	-0.010
10.0	44.969	2.248	47.187	-0.986

Table 2.3.3   Data generated with Eq. 2.3.6, error in Y_5, fit with $y = a_1 + a_2x + a_3x^2$. Rel_Error is $(Y - y) / \sigma_y$.

We can explore the effect of **not** using Equation 2.3.3 when the values of σ_y vary significantly from point to point. The data used to generate Figure 2.3.1 are shown in Table 2.3.4. Note the large relative errors for the first few points when unit weighting (i.e., $w_i = 1$) is used.

x	Y	σ_y	$(Y-y)/\sigma_y$ $(w_i=1)$	$(Y-y)/\sigma_y$ $(w_i=1/\sigma_y^2)$
1	6.90	0.05	15.86	0.19
2	11.95	0.10	3.14	-0.39
3	16.800	0.20	-1.82	-0.44
4	22.500	0.50	-0.38	1.23
5	26.200	0.80	-2.53	-0.85
6	33.500	1.50	-0.17	1.08
7	41.000	4.00	0.43	1.03

Table 2.3.4   Fitting Fig 2.3.1 data using $y = a_1 + a_2x$: different weighting schemes.

In the discussion preceding Equation 2.3.3 it was assumed that the errors in the independent variable (or variables) were negligible. If this assumption cannot be made, then if we assume that the noise component in the data is relatively small, it can be shown [WO67] that the following equation can be used instead of 2.3.3 for the weights w_i:

$$w_i = \frac{1}{\sigma_{y_i}^2 + \sum\limits_{j=1}^{j=m} \left(\frac{\partial f}{\partial x_j} \sigma_{x_{j_i}} \right)^2} \tag{2.3.7}$$

This equation is a more generalized form of **statistical weighting** than Equation 2.3.3. The derivation of this equation is based upon the assumption that higher order terms can be neglected in a Taylor expansion in the region near the minimum value of S. As an example of the application of 2.3.7 to a specific problem, the weighting function for the parabolic fit would be the following if the σ_x's are included in the analysis:

$$w_i = \frac{1}{\sigma_{y_i}^2 + ((a_2 + 2a_3 x_i)\sigma_{x_i})^2}$$

In this equation, since there is only one independent variable, we can eliminate the subscript j and use only x rather than x_1. As a 2nd example, consider the following function:

$$y = a_1 + a_2 x_1 + a_2 x_2$$

This equation has two independent variables and the weights w_i would be computed as follows:

$$w_i = \frac{1}{\sigma_{y_i}^2 + (a_2 \sigma_{x_{1_i}})^2 + (a_3 \sigma_{x_{2_i}})^2}$$

2.4 Obtaining the Least Squares Solution

The least squares solution is defined as the point in the "unknown parameter" space at which the objective function S is minimized. Thus, if there are p unknown parameters (a_k, $k = 1$ to p), the solution yields the values of the a_k's that minimize S. To find this minimum point we set the p partial derivatives of S to zero yielding p equations for the p unknown values of a_k:

$$\frac{\partial S}{\partial a_k} = 0 \qquad k = 1 \text{ to } p \tag{2.4.1}$$

In Section 2.2 the following expression (Equation 2.2.3) for the objective function S was developed:

$$S = \sum_{i=1}^{i=n} w_i (Y_i - f(X_i))^2$$

In this expression the independent variable X_i can be either a scalar or a vector. The variable Y_i can also be a vector but is usually a scalar. Using this expression and Equation 2.4.1, we get the following p equations:

$$-2 \sum_{i=1}^{i=n} w_i (Y_i - f(X_i)) \frac{\partial f(X_i)}{\partial a_k} = 0 \qquad k = 1 \text{ to } p$$

$$\sum_{i=1}^{i=n} w_i f(X_i) \frac{\partial f(X_i)}{\partial a_k} = \sum_{i=1}^{i=n} w_i Y_i \frac{\partial f(X_i)}{\partial a_k} \qquad k = 1 \text{ to } p \tag{2.4.2}$$

For problems in which the function f is linear, Equation 2.4.2 can be solved directly. In Section 2.1 Equation 2.1.3 was used to specify linear equations:

$$y = f(X) = \sum_{i=1}^{i=p} a_k g_k(X) = \sum_{i=1}^{i=p} a_k g_k(x_1, x_2 .... x_m)$$

The derivatives of f are simply:

$$\frac{\partial f(X)}{\partial a_k} = g_k(X) \qquad k = 1 \text{ to } p \tag{2.4.3}$$

Substituting 2.4.3 into 2.4.2 we get the following set of equations:

$$\sum_{j=1}^{j=p} a_j \sum_{i=1}^{i=n} w_i g_j(\mathbf{X}_i) g_k(\mathbf{X}_i) = \sum_{i=1}^{i=n} w_i Y_i g_k(\mathbf{X}_i) \quad k=1 \text{ to } p \qquad (2.4.4)$$

Simplifying the notation by using g_k instead of $g_k(X_i)$ we get the following set of equations:

$$a_1 \sum w_i g_1 g_k + a_2 \sum w_i g_2 g_k + \dots + a_p \sum w_i g_p g_k = \sum w_i Y_i g_k \qquad (2.4.5)$$

$$k = 1 \text{ to } p$$

We can rewrite these equations using matrix notation:

$$CA = V \qquad (2.4.6)$$

In this equation C is a p by p matrix and A and V are vectors of length p. The terms C_{jk} and V_k are computed as follows:

$$C_{jk} = \sum_{i=1}^{i=n} w_i g_j g_k \qquad (2.4.7)$$

$$V_k = \sum_{i=1}^{i=n} w_i Y_i g_k \qquad (2.4.8)$$

The terms of the A vector (i.e., the unknown parameters a_k) are computed by solving the matrix equation 2.4.6:

$$A = C^{-1}V \qquad (2.4.9)$$

In this equation, C^{-1} is the inverse matrix of C. As an example, let us consider problems in which a straight line is fit to the data:

$$y = f(x) = a_1 + a_2 x \qquad (2.4.10)$$

For this equation $g_1 = 1$ and $g_2 = x$ so the C matrix is:

$$C = \begin{bmatrix} \sum\limits_{i=1}^{i=n} w_i & \sum\limits_{i=1}^{i=n} w_i x_i \\[2ex] \sum\limits_{i=1}^{i=n} w_i x_i & \sum\limits_{i=1}^{i=n} w_i x_i^2 \end{bmatrix} \tag{2.4.11}$$

The V vector is:

$$V = \begin{bmatrix} \sum\limits_{i=1}^{i=n} w_i Y_i \\[2ex] \sum\limits_{i=1}^{i=n} w_i Y_i x_i \end{bmatrix} \tag{2.4.12}$$

To apply these equations to a real set of data, let us use the 7 points included in Table 2.3.4 and let us use the case in which all the values of w_i are set to 1 (i.e., unit weighting). For this case, the C and C^{-1} matrices and the V vector are:

$$C = \begin{bmatrix} 7 & 28 \\ 28 & 140 \end{bmatrix} \quad C^{-1} = \frac{1}{196}\begin{bmatrix} 140 & -28 \\ -28 & 7 \end{bmatrix} \quad V = \begin{bmatrix} 158.85 \\ 790.20 \end{bmatrix}$$

Solving Equation 2.4.9:

$$A = C^{-1}V = \begin{bmatrix} C_{11}^{-1}V_1 + C_{12}^{-1}V_2 \\ C_{21}^{-1}V_1 + C_{22}^{-1}V_2 \end{bmatrix} = \begin{bmatrix} 0.5786 \\ 5.5286 \end{bmatrix} \tag{2.4.13}$$

For problems in which the f function is nonlinear, the procedure is similar but is iterative. One starts with initial guesses $a0_k$ for the unknown values of a_k. Simplifying the notation used for Equation 2.4.2, we see that the equations for terms of the C matrix and V vector are:

$$C_{jk} = \sum\limits_{i=1}^{i=n} w_i \frac{\partial f}{\partial a_j} \frac{\partial f}{\partial a_k} \tag{2.4.14}$$

$$V_k = \sum\limits_{i=1}^{i=n} w_i Y_i \frac{\partial f}{\partial a_k} \tag{2.4.15}$$

In the equation for V_k the parameter Y_i is no longer the value of the dependent variable. It is value of the dependent variable minus the computed values using the initial guesses (i.e., the $a0_k$'s). For linear problems we don't need to make this distinction because the initial guesses are zero and thus the computed values are zero. The A vector is determined using Equation 2.4.9 but for nonlinear problems, this vector is no longer the solution vector. It is the vector of computed changes in values of the initial guesses $a0_k$:

$$a_k = a0_k + A_k \qquad k = 1 \text{ to } p \tag{2.4.16}$$

The values of a_k are then used as initial guesses $a0_k$ for the next iteration. This process is continued until a convergence criterion is met or the process does not achieve convergence. Typically the convergence criterion requires that the fractional changes are all less than some specified value of ε.

$$\left| A_k / a0_k \right| \leq \varepsilon \qquad k = 1 \text{ to } p \tag{2.4.17}$$

Clearly this convergence criterion must be modified if a value of $a0_k$ is zero or very close to zero. For such terms, one would only test the absolute value of A_k and not the relative value. This method of converging towards a solution is called the Gauss-Newton algorithm and will lead to convergence for many nonlinear problems [WO67]. It is not, however, the only search algorithm and a number of alternatives to Gauss-Newton are discussed in Section 6.4.

As an example, of a nonlinear problem, let us once again use the data in Table 2.3.4 but choose the following nonlinear exponential function for f:

$$y = f(x) = a_1 e^{a_2 x} \tag{2.4.18}$$

The two derivatives of this function are:

$$f_1' \equiv \frac{\partial f}{\partial a_1} = e^{a_2 x} \qquad \text{and} \qquad f_2' \equiv \frac{\partial f}{\partial a_2} = x a_1 e^{a_2 x}$$

Let us choose as initial guesses $a0_1=1$ and $a0_2=0.1$ and weights $w_i=1$. (Note that if an initial guess of $a0_1=0$ is chosen, all values of the derivative of f with respect to a_2 will be zero. Thus all the terms of the C matrix except C_{11} will be zero. The C matrix would then be singular and no solution could be obtained for the A vector.) Using Equations 2.4.14 and 2.4.15 and the expressions for the derivatives, we can compute the terms of the C matrix and V vector and then using 2.4.9 we can solve for the values of A_1 and A_2. The computed values are 4.3750 and 2.1706 therefore the initial values for the next iteration are 5.3750 and 2.2706. Using Equation 2.4.17 as the convergence criterion and a value of $\varepsilon = 0.001$, final values of $a_1 = 7.7453$ and $a_2 = 0.2416$ are obtained. The value of S obtained using the initial guesses is approximately 1,260,000. The value obtained using the final values of a_1 and a_2 is 17.61. Details of the calculation for the first iteration are included in Tables 2.4.1 and 2.4.2.

x	y	$f=1.0e^{0.1x}$	$Y=y-f$
1	6.900	1.105	5.795
2	11.950	1.221	10.729
3	16.800	1.350	15.450
4	22.500	1.492	21.008
5	26.200	1.649	24.551
6	33.500	1.822	31.678
7	41.000	2.014	38.986

Table 2.4.1   Fitting Data using $f(x)=a_1e^{a_2x}$ **with initial guesses** $a_1=1$, $a_2=0.1$

Point	$(f_1')^2$	$(f_2')^2$	$f_1'f_2'$	$f_1'Y$	$f_2'Y$
1	1.221	1.221	1.221	6.404	6.404
2	1.492	5.967	2.984	13.104	26.209
3	1.822	16.399	5.466	20.855	62.565
4	2.226	35.609	8.902	31.340	125.360
5	2.718	67.957	13.591	40.478	202.390
6	3.320	119.524	19.921	57.721	346.325
7	4.055	198.705	28.386	78.508	549.556
Sum	16.854	445.382	80.472	248.412	1318.820
	C_{11}	C_{22}	C_{12}	V_1	V_2

Table 2.4.2   Computing terms of the C matrix and V vector

Using the values of the terms of the C matrix and V vector from Table 2.4.2, and solving for the terms of the A vector using Equation 2.4.9, we get values of $a_1 = 4.3750$ and $a_2 = 2.1706$. Using Equation 2.4.16, the values of the initial guesses for the next iteration are therefore 5.3750 and 2.2706. This process is repeated until convergence is obtained. As the initial guesses improve from iteration to iteration, the computed values of the dependent variable (i.e., f) become closer to the actual values of the dependent variable (i.e., y) and therefore the differences (i.e., Y) become closer to zero. From Equation 2.4.15 we see that the values of V_k become smaller as the process progresses towards convergence and thus the terms of the A vector become smaller until the convergence criterion (Equation 2.4.17) is achieved.

A question sometimes asked is: if we increase or decrease the weights how does this affect the results? For example, for unit weighting what happens if we use a value of w other than 1? The answer is that it makes no difference. The values of the terms of the V vector will be proportional to w and all the terms of the C matrix will also be proportional to w. The C^{-1} matrix, however, will be inversely proportional to w and therefore the terms of the A vector (i.e., the product of $C^{-1}V$) will be independent of w. A similar argument can be made for statistical weighting. For example, if all the values of σ_y are increased by a factor of 10, the values of w_i will be decreased by a factor of 100. Thus all the terms of V and C will be decreased by a factor of 100, the terms of C^{-1} will be increased by a factor of 100 and the terms of A will remain unchanged. What makes a difference are the relative values of the weights and not the absolute values. We will see, however, when Goodness-of-Fit is discussed in Chapter 3, that an estimate of the amplitude of the noise component of the data can be very helpful. Furthermore, if prior estimates of the unknown parameters of the model are included in the analysis, then the weights of the data points must be based upon estimates of the absolute values of the weights.

2.5 Uncertainty in the Model Parameters

In Section 2.4 we developed the methodology for finding the set of a_k's that minimize the objective function S. In this section we turn to the task of determining the uncertainties associated with the a_k's. The usual measures of uncertainty are standard deviation (i.e., σ) or variance (i.e., σ^2) so we seek an expression that allows us to estimate the σ_{a_k}'s. It can be shown [WO67, BA74, GA92] that the following expression gives us an un-biased estimate of σ_{a_k}:

$$\sigma_{a_k}^2 = \frac{S}{n-p} C_{kk}^{-1}$$

$$\sigma_{a_k} = (\frac{S}{n-p} C_{kk}^{-1})^{1/2} \qquad (2.5.1)$$

We see from this equation that the unbiased estimate of σ_{a_k} is related to the objective function S and the kth diagonal term of the inverse matrix C^{-1}. The matrix C^{-1} is required to find the least squares values of the a_k's and once these values have been determined, the final (i.e., minimum) value of S can easily be computed. Thus the process of determining the a_k's leads painlessly to a determination of the σ_{a_k}'s.

As an example, consider the data included in Table 2.3.4. In Section 2.4 details were included for a straight-line fit to the data using unit weighting:

$$y = f(x) = a_1 + a_2 x = 0.5786 + 5.5286x \qquad (2.5.2)$$

The C and C^{-1} matrices were:

$$C = \begin{bmatrix} 7 & 28 \\ 28 & 140 \end{bmatrix} \qquad C^{-1} = \frac{1}{196} \begin{bmatrix} 140 & -28 \\ -28 & 7 \end{bmatrix}$$

The value for $S/(n-p) = S/(7-2)$ is 1.6019. We can compute the σ_{a_k}'s from Equation 2.5.1:

$$\sigma_{a_1} = \sqrt{1.6019 * 140/196} = 1.070 \quad \text{and} \quad \sigma_{a_2} = \sqrt{1.6019 * 7/196} = 0.2392$$

The relative error in a_1 is $1.070 / 0.5786 = 1.85$ and the relative error in a_2 is $0.2392 / 5.5286 = 0.043$. If the purpose of the experiment was to determine, a_2, then we have done fairly well (i.e., we have determined a_2 to about 4%). However, if the purpose of the experiment was to determine, a_1, then we have done terribly (i.e., the relative error is about 185%).

What does this large relative error imply? If we were to repeat the experiment many times, we would expect that the computed value of a_1 would fall within the range $0.5786 \pm 2.57 * 1.85 = 0.5786 \pm 4.75$ ninety-five percent of the time (i.e., from -4.17 to 5.33). This is a very large range of probable results. (The constant 2.57 comes from the t distribution with 5 degrees of freedom.)

If we use statistical weighting (i.e., $w_i=1/\sigma_y^2$), can we improve upon these results? Reanalyzing the data in Table 2.3.4 using the values of σ_y included in the table, we get the following straight line:

$$y = f(x) = a_1 + a_2 x = 1.8926 + 4.9982x \qquad (2.5.3)$$

The computed value of σ_{a_1} is 0.0976 and the value for σ_{a_2} is 0.0664. These values are considerably less than the values obtained using unit weighting. The reduction in the value of σ_{a_1} is more than a factor of 10 and the reduction in σ_{a_2} is almost a factor of 4. This improvement in the accuracy of the results is due to the fact that in addition to the actual data (i.e., the values of x and y) the quality of the data (i.e., σ_y) was also taken into consideration.

We should also question the independence of a_1 and a_2. If for example, we repeat the experiment many times and determine many pairs of values for a_1 and a_2, how should the points be distributed in the two-dimensional space defined by a_1 and a_2? Are they randomly scattered about the point $[a_1 = 0.5786, a_2 = 5.5286]$ or is there some sort of correlation between these two parameters? An answer to this question is also found in the least squares formulation. The notation σ_{jk} is used for the **covariance** between the parameters j and k and is computed as follows:

$$\sigma_{jk} = \frac{S}{n-p} C_{jk}^{-1} \qquad (2.5.4)$$

A more meaningful parameter is the **correlation coefficient** between the parameters j and k. Denoting this parameter as ρ_{jk}, we compute it as follows:

$$\rho_{jk} = \frac{\sigma_{jk}}{\sigma_{a_j}\sigma_{a_k}} \qquad (2.5.5)$$

The correlation coefficient is a measure of the degree of correlation between the parameters. The values of ρ_{jk} are in the range from -1 to 1. If
the value is zero, then the parameters are uncorrelated (i.e., independent),
if the value is 1, then they fall exactly on a line with a positive slope and if
the value is -1 then the fall exactly on a line with a negative slope. Examples of different values of ρ_{jk} are seen in Figure 2.5.1.

Returning to our example using unit weighting, let us compute σ_{12} and
ρ_{12}:

$$\sigma_{12} = -1.6019 * 28/196 = -0.2288$$

$$\rho_{12} = \frac{-0.2288}{1.070 * 0.2392} = -0.894$$

In other words, a_1 and a_2 are strongly negatively correlated. Larger-than-
average values of a_1 are typically paired with smaller-than-average values
of a_2.

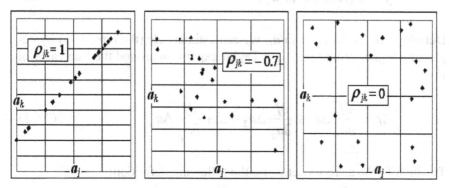

**Figure 2.5.1   Correlation Coefficients for Several Different Data
Distributions**

As will be seen in Section 2.6, the covariance is used in evaluating the
standard deviations of the least squares curves. For example, we can use
Equation 2.5.2 or 2.5.3 to predict the value of y for any value of x. The

covariance is needed to estimate the uncertainty σ_f associated with the predicted value of y (i.e., $f(\mathbf{X})$).

2.6 Uncertainty in the Model Predictions

In Section 2.5 the uncertainties in the model parameters were considered. If the only purpose of the experiment is to determine the parameters of the model, then only these uncertainties are of interest. However, there are many situations in which we are interested in using the model for making predictions. Once the parameters of the model are available, then the equation $f(\mathbf{X})$ can be used to predict y for any combination of the independent variables (i.e., the vector $\mathbf{X}$). In this section attention is turned towards the uncertainties σ_f of these predictions.

Typically, one assumes that the model is "correct" and thus the computed values of y are normally distributed about the true values. For a given set of values for the terms of the $\mathbf{X}$ vector (i.e., a combination of the independent variables $x_1, x_2,.., x_m$), we assume that the uncertainty in the predicted value of y is due to the uncertainties associated with the a_k's. The predicted value of y is determined by substituting $\mathbf{X}$ into $f(\mathbf{X})$:

$$y = f(\mathbf{X}; a_1, a_2,..., a_p) \tag{2.6.1}$$

Defining Δa_k as the error in a_k, we can estimate Δy (the error in y) by neglecting higher order terms in a Taylor expansion around the true value of y:

$$\Delta f \cong \frac{\partial f}{\partial a_1} \Delta a_1 + \frac{\partial f}{\partial a_2} \Delta a_2 + ... + \frac{\partial f}{\partial a_p} \Delta a_p \tag{2.6.2}$$

To simplify the analysis, let is use the following definition:

$$T_k = \frac{\partial f}{\partial a_k} \Delta a_k \tag{2.6.3}$$

Thus:

$$\Delta f \cong T_1 + T_2 + ... + T_p = \sum_{k=1}^{k=p} T_k \qquad (2.6.4)$$

If we square Equation 2.6.2 we get the following:

$$\Delta f^2 \cong (T_1)^2 + (T_2)^2 .. + (T_p)^2 + 2T_1T_2 + 2T_1T_3 .. + 2T_{p-1}T_p \qquad (2.6.5)$$

$$\Delta f^2 \cong \sum_{k=1}^{k=p} (T_k)^2 + \sum_{j=1}^{j=p} \sum_{k=j+1}^{k=p} 2T_jT_k \qquad (2.6.6)$$

If the experiment is repeated many times and average values of the terms are taken, we obtain the following from Equation 2.6.6:

$$(\Delta f^2)_{avg} \cong \sum_{k=1}^{k=p} ((T_k)^2)_{avg} + \sum_{j=1}^{j=p} \sum_{k=j+1}^{k=p} 2(T_jT_k)_{avg} \qquad (2.6.7)$$

Recognizing that $((\Delta a_k)^2)_{avg}$ is just $(\sigma_{a_k})^2$ and $(\Delta a_j \Delta a_k)_{avg}$ is σ_{jk} we get the following:

$$\sigma_f^2 = \sum_{k=1}^{k=p} (\frac{\partial f}{\partial a_k} \sigma_{a_k})^2 + \sum_{j=1}^{j=p} \sum_{k=j+1}^{k=p} 2 \frac{\partial f}{\partial a_j} \frac{\partial f}{\partial a_k} \sigma_{jk} \qquad (2.6.8)$$

The number of cross-product terms (i.e., terms containing σ_{jk}) is $p\,(p-1)\,/\,2$. If we use the following substitution:

$$\sigma_{kk} = \sigma_{a_k}^2 \qquad (2.6.9)$$

and recognizing that $\sigma_{jk} = \sigma_{kj}$ we can simplify equation 2.6.8:

$$\sigma_f^2 = \sum_{j=1}^{j=p} \sum_{k=1}^{k=p} \frac{\partial f}{\partial a_j} \frac{\partial f}{\partial a_k} \sigma_{jk} \qquad (2.6.10)$$

Using Equations 2.5.1 and 2.5.4, we can relate σ_f to the terms of the C^{-1} matrix:

$$\sigma_f^2 = \frac{S}{n-p} \sum_{j=1}^{j=p} \sum_{k=1}^{k=p} \frac{\partial f}{\partial a_j} \frac{\partial f}{\partial a_k} C_{jk}^{-1} \qquad\qquad (2.6.11)$$

As an example of the application of this equation to a data set, let us once again use the data from Table 2.3.4 and $w_i = 1$. The data was fit using a straight line:

$$y = f(x) = a_1 + a_2 x$$

so the derivatives are:

$$\frac{\partial f}{\partial a_1} = 1 \qquad \text{and} \qquad \frac{\partial f}{\partial a_2} = x$$

We have seen that the inverse matrix is:

$$C^{-1} = \frac{1}{196} \begin{bmatrix} 140 & -28 \\ -28 & 7 \end{bmatrix}$$

and the value of $S/(n-p)$ is 1.6019. Substituting all this into Equation 2.6.11 we get the following expression:

$$\sigma_f^2 = \frac{S}{n-p} \left(\frac{\partial f}{\partial a_1} \frac{\partial f}{\partial a_1} C_{11}^{-1} + \frac{\partial f}{\partial a_2} \frac{\partial f}{\partial a_2} C_{22}^{-1} + 2 \frac{\partial f}{\partial a_1} \frac{\partial f}{\partial a_2} C_{12}^{-1} \right)$$

$$\sigma_f^2 = \frac{S}{n-p} (C_{11}^{-1} + x^2 C_{22}^{-1} + 2x C_{12}^{-1})$$

$$\sigma_f^2 = \frac{1.6019}{196} (140 + 7x^2 - 56x) \qquad\qquad (2.6.12)$$

(Note that the C^{-1} matrix is always symmetric so we can use $2C_{jk}^{-1}$ instead of $C_{jk}^{-1} + C_{kj}^{-1}$.)

Equations 2.5.2 and 2.6.12 are used to predict values of y and σ_f for several values of x and the results are seen in Table 2.6.1. Note the curious fact that the values of σ_f are symmetric about $x = 4$. This phenomenon is easily explained by examining Equation 2.6.12 and noting that this equation is a parabola with a minimum value at $x = 4$.

x	$y=f(x)$	σ_f
1.5	8.871	0.766
2.5	14.400	0.598
3.5	19.929	0.493
4.5	25.457	0.493
5.5	30.986	0.598
6.5	36.514	0.766

Table 2.6.1   Predicted values of y and σ_f using $w_i=1$.

In the table, we see that the values of x that have been chosen are all within the range of the values of x used to obtain the model (i.e., 1 to 7). The use of a model for purposes of extrapolation should be done with extreme caution! (More will be said about extrapolation in Chapter 3.) Note that the σ_f values tend to be least at the midpoint of the range and greatest at the extreme points. This is reasonable. Instinctively if all the data points are weighted equally, we would expect σ_f to be least in regions that are surrounded by many points. Table 2.6.1 was based upon a least squares analysis in which all points were weighted equally. However, when the points are not weighted equally, the results can be quite different. Table 2.6.2 is also based upon the x and y values from Table 2.3.4 but using statistical weighting (i.e., $w_i=1/\sigma_y^2$).

Table 2.6.2 presents a very different picture than Table 2.6.1 (which is based upon unit weighting). When unit weighting is used differences in the quality of the data are ignored, and we see (in Table 2.3.4) that the relative errors for the first few data points are large. However, when the data is statistically weighted, the relative errors (also seen in Table 2.3.4) are all comparable. In Table 2.6.2 we observe that the values of σ_f are much less than the values in Table 2.6.1 even for points at the upper end of the range. This improvement in accuracy is a result of taking the quality of the data into consideration (i.e., using statistical weighting). Furthermore, the most accurate points (i.e., least values of σ_f) are near the points that have the smallest values of σ_y.

x	$y=f(x)$	σ_f
1.5	9.390	0.044
2.5	14.388	0.089
3.5	19.386	0.151
4.5	24.384	0.215
5.5	29.383	0.281
6.5	34.381	0.347

Table 2.6.2   Predicted values of y and σ_f using $w_i=1/\sigma_y^2$.

In Section 2.4 we noted that increasing or decreasing weight by a constant factor had no effect upon the results (i.e., the resulting A vector). Similarly, changes in the weights do not affect the computed values of σ_f and σ_{a_k}. The value of S and the terms of the C matrix will change proportionally if the w's are changed by a constant factor and the changes in the terms of the C^{-1} matrix will be inversely proportional to the changes in w. The computation of both σ_f and σ_{a_k} are based upon the products of S and terms of the C^{-1} matrix so they will be independent of proportional changes in w. What does makes a difference is the relative values of the weights and not the absolute values. This does not imply that estimates of the actual rather than the relative uncertainties of the data are unimportant. When Goodness-of-Fit is discussed in Chapter 3, we will see that an estimate of the amplitude of the noise component of the data can be very helpful.

It should be emphasized that values of σ_f computed using Equation 2.6.11 are the σ's associated with the function f and are a measure of how close the least squares curve is to the "true" curve. One would expect that as the number of points increases, the values of σ_f decreases and if the function f is truly representative of the data σ_f will approach zero as n approaches infinity. Equation 2.6.11 does in fact lead to this conclusion. The term $S / (n-p)$ approaches one and the terms of the C matrix become increasingly large for large values of n. The terms of the C^{-1} matrix therefore become increasingly small and approach zero in the limit of n approaching infinity. In fact one can draw a "95% confidence band" around the computed function f. The interpretation of this band is that for a given value of x the probability that the "true" value of f falls within these limits is 95%. Sometimes we are more interested in the "95% prediction band". Within this band we would expect that 95% of new data points will fall [MO03].

This band is definitely not the same as the 95% confidence band and the effect of increasing n has only a small effect upon the prediction band. Assuming that for a given x the deviations from the true curve and from the least squares curve are independent, the σ's associated with the prediction band are computed as follows:

$$\sigma^2_{pred} = \sigma^2_f + \sigma^2_y \qquad (2.6.13)$$

Knowing that as n increases, σ_f becomes increasingly small, the limiting value of σ_{pred} is σ_y. In Table 2.6.3 the values of σ_{pred} are computed for the same data as used in Table 2.6.2. The values of σ_y are interpolated from the values in Table 2.3.4. The 95% prediction band is computed using σ_{pred} and the value of t corresponding to 95% limits for $n - p$ degrees of freedom. From Table 2.3.4 we see that $n = 7$ and for the straight line fit p $=2$. The value of t for $\alpha = 2.5\%$ and 5 degrees of freedom is 2.571. In other words, 2.5% of new points should fall above $f(x) + 2.571\sigma_{pred}$ and 2.5% should fall below $f(x) - 2.571\sigma_{pred}$. The remaining 95% should fall within this band. As n increases, the value of t approaches the value for the standard normal distribution which for a 95% confidence limit is 1.96. The 95% confidence and prediction bands for this data are seen in Figure 2.6.1.

x	$y=f(x)$	σ_y	σ_f	σ_{pred}
1.5	9.390	0.075	0.044	0.087
2.5	14.388	0.150	0.089	0.174
3.5	19.386	0.350	0.151	0.381
4.5	24.384	0.650	0.215	0.685
5.5	29.383	1.150	0.281	1.184
6.5	34.381	2.750	0.347	2.772

Table 2.6.3   Values of σ_{pred} using data from Table 2.3.4 and statistical weighting

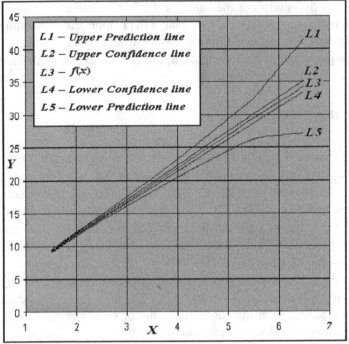

Figure 2.6.1   Confidence and Prediction Bands for Data from Table 2.6.3

2.7 Treatment of Prior Estimates

In the previous sections we noted that a basic requirement of the method of least squares is that the number of data points n must exceed p (the number of unknown parameters of the model). The difference between these two numbers $n-p$ is called the "number of degrees of freedom". Very early in my career I came across an experiment in which the value of $n-p$ was in fact negative! The modeling effort was related to damage caused by a certain type of event and data had been obtained based upon only two events. Yet the model included over ten unknown parameters. The independent variables included the power of the event and other variables related to position. To make up the deficit, estimates of the parameters based upon theoretical models were used to supplement the two data points. The prior estimates of the parameters are called Bayesian estimators and if the number of Bayesian estimators is n_b then the number of degrees of freedom is $n+n_b-p$. As long as this number is greater than zero, a least squares calculation can be made.

In Section 2.2 Equation 2.2.6 is the modified form that the objective function takes when prior estimates of the a_k parameters are available:

$$S = \sum_{i=1}^{i=n} w_i (Y_i - f(X_i))^2 + \sum_{k=1}^{k=p} (a_k - b_k)^2 / \sigma_{b_k}^2$$

In this equation b_k is the prior estimates of a_k and σ_{b_k} is the uncertainty associated with this prior estimate. The parameter b_k is typically used as the initial guess $a0_k$ for a_k. We see from this equation that each value of b_k is treated as an additional data point. However, if σ_{b_k} is not specified, then it is assumed to be infinite and no weight is associated with this point. In other words, if σ_{b_k} is not specified then b_k is treated as just an initial guess for a_k and not as a prior estimate. The number of values of b_k that are specified (i.e., not infinity) is n_b.

In the previous sections it was stated that the weights w_i could be based upon relative and not absolute values of the uncertainties associated with the data points. When prior estimates of the a_k's are included in the analysis, we are no longer at liberty to use relative weights. Since the weights associated with the prior estimates are based upon estimates of absolute values (i.e., $1 / (\sigma_{b_k})^2$), the w_i values must also be based upon estimates of absolute values.

To find the least squares solution, we proceed as in Section 2.4 by setting the p partial derivatives of S to zero yielding p equations for the p unknown values of a_k:

$$-2 \sum_{i=1}^{i=n} w_i (Y_i - f(X_i)) \frac{\partial f(X_i)}{\partial a_k} + 2 \sum_{k=1}^{k=p} (a_k - b_k) / \sigma_{b_k}^2 = 0 \qquad k = 1 \text{ to } p$$

The terms in the last summation can be expanded:

$$(a_k - b_k)/\sigma_{b_k}^2 = (a_k - a0_k + a0_k - b_k)/\sigma_{b_k}^2 = A_k/\sigma_{b_k}^2 + (a0_k - b_k)/\sigma_{b_k}^2$$

$$\sum_{i=1}^{i=n} w_i f(\mathbf{X}_i) \frac{\partial f(\mathbf{X}_i)}{\partial a_k} + \sum_{k=1}^{k=p} \frac{A_k}{\sigma_{b_k}^2} = \sum_{i=1}^{i=n} w_i Y_i \frac{\partial f(\mathbf{X}_i)}{\partial a_k} + \sum_{k=1}^{k=p} \frac{b_k - a0_k}{\sigma_{b_k}^2} \qquad (2.7.1)$$

As in Section 2.4 this equation is best treated as a matrix equation:

$$CA = V$$

The diagonal terms of the C matrix are modified but the off-diagonal terms remain the same:

$$C_{jk} = C_{kj} = \sum_{i=1}^{i=n} w_i \frac{\partial f}{\partial a_j} \frac{\partial f}{\partial a_k} \qquad (j \neq k) \qquad (2.7.2)$$

$$C_{kk} = \frac{1}{\sigma_{b_k}^2} + \sum_{i=1}^{i=n} w_i \frac{\partial f}{\partial a_j} \frac{\partial f}{\partial a_k} \qquad (2.7.3)$$

The terms of the V vector are also modified:

$$V_k = \frac{b_k - a0_k}{\sigma_{b_k}^2} + \sum_{i=1}^{i=n} w_i Y_i \frac{\partial f}{\partial a_k} \qquad (2.7.4)$$

Solution of the matrix equation $CA = V$ yields the vector A which is then used to compute the unknown a_k's (Equation 2.4.16). The computation of the σ_{a_k} terms must be modified to include the additional data points. The modified form of Equation 2.5.1 is:

$$\sigma_{a_k} = \left(\frac{S}{n + n_b - p} C_{kk}^{-1} \right)^{1/2} \qquad (2.7.5)$$

In this equation n_b is the number of Bayesian estimations included in the analysis (i.e., the number of b_k's that are specified). Using the same reasoning, Equation 2.6.11 must also be modified:

$$\sigma_f^2 = \frac{S}{n + n_b - p} \sum_{j=1}^{j=p} \sum_{k=1}^{k=p} \frac{\partial f}{\partial a_j} \frac{\partial f}{\partial a_k} C_{jk}^{-1} \tag{2.7.6}$$

As an example of the application of prior estimates, let us once again use the data in Table 2.3.4 but only for the case of statistical weighting (i.e., $w = 1/\sigma_y^2$). The straight line computed for this case was:

$$y = f(x) = a_1 + a_2 x = 1.8926 + 4.9982x$$

The computed value of σ_{a_1} and σ_{a_2} were 0.0976 and 0.0664. The C matrix and V vector for this case are:

$$C = \begin{bmatrix} 531.069 & 701.917 \\ 701.917 & 1147.125 \end{bmatrix} \quad \text{and} \quad V = \begin{bmatrix} 4513.39 \\ 7016.96 \end{bmatrix}$$

Let us say that we have a prior estimate of a_1:

$$b_1 = 1.00 \pm 0.10$$

The only term in the C matrix that changes is C_{11}. The terms of the V vector are, however, affected by the changes in the values of Y_i. Since we start from the initial guess for a_1, all the values of Y_i are reduced by a_1 (i.e., 1.00) :

$$C = \begin{bmatrix} 631.069 & 701.917 \\ 701.917 & 1147.125 \end{bmatrix} \quad \text{and} \quad V = \begin{bmatrix} 3982.32 \\ 6360.04 \end{bmatrix}$$

Solving for the terms of the A vector we get $A_1 = 0.4498$ and $A_2 = 5.2691$. The computed value of a_1 is therefore 1.4498 and a_2 is 5.2691. Note that the prior estimate of a_1 reduces the value previously computed from 1.8926 towards the prior estimate of 1.00. The values of σ_{a_1} and σ_{a_2} for this calculation were 0.1929 and 0.1431. These values are greater than the values obtained without the prior estimate and that indicates that the prior estimate of a_1 is not in agreement with the experimental results. Assuming that there is no discrepancy between the prior estimates and the experimental data, we would expect a reduction in uncertainty. For example, if we repeat the analysis but use as our prior estimate: $b_1 = 2.00 \pm 0.10$:

The resulting values of a_1 and a_2 are:

$$a_1 = 1.9459 \pm 0.0670 \qquad a_2 = 4.9656 \pm 0.0497$$

If we repeat the analysis and use prior estimates for both a_1 and a_2:

$$b_1 = 2.00 \pm 0.10 \qquad b_2 = 5.00 \pm 0.05$$

The resulting values of a_1 and a_2 are:

$$a_1 = 1.9259 \pm 0.0508 \qquad a_2 = 4.9835 \pm 0.0325$$

The results for all these cases are summarized in Table 2.7.1.

b_1	b_2	$n+n_b$	a_1	σ_{a_1}	a_2	σ_{a_2}
none	None	7	1.8926	0.0976	4.9982	0.0664
1.00±0.1	None	8	1.4498	0.1929	5.2691	0.1431
2.00±0.1	None	8	1.9459	0.0670	4.9656	0.0497
2.00±0.1	5.00±0.05	9	1.9259	0.0508	4.9835	0.0325

Table 2.7.1   Computed values of a_1 and a_2 for combinations of b_1 and b_2.

We see in this table that the best results (i.e., minimum values of the σ's) are achieved when the prior estimates are in close agreement with the results obtained without the benefit of prior estimates of the unknown parameters a_1 and a_2.

2.8 Applying Least Squares to Classification Problems

In the previous sections the dependent variable y was assumed to be a continuous numerical variable and the method of least squares was used to develop models that could then be used to predict the value of y for any combination of the independent x variable (or variables). There are, however, problems in which the dependent variable is a "class" rather than a continuous variable. For example the problem might require a model that differentiates between two classes: "good" or "bad" or three levels: "low", "medium" or "high". Typically we have **nlrn** learning points that can be used to create the model and then **ntst** test points that can be used to test how well the model predicts on unseen data. The method of least squares can be applied to classification problems in a very straight-forward manner.

The trick that allows a very simple least squares solution to classification problems is to assign numerical values to the classes (i.e., the y values) and then make predictions based upon the computed value of y for each test point. For example, for two class problems we can assign the values 0 and 1 to the two classes (e.g., "bad" = 0 and "good" = 1). We then fit the learning data using least squares as the modeling technique and then for any combination of the x variables, we compute the value of y. If it is less than 0.5 the test point is assumed to fall within the "bad" class, otherwise it is classified as "good". For 3 class problems we might assign 0 to class 1, 0.5 to class 2 and 1 to class 3. If a predicted value of y is less than 1/3 then we would assign class 1 as our prediction, else if the value was less than 2/3 we would assign class 2 as our prediction, otherwise the assignment would be class 3. Obviously the same logic can be applied to any number of classes.

It should be emphasized that the least squares criterion is only one of many that can be used for classification problems.   In their book on *Statistical Learning*, Hastie, Tibshirani and Friedman discuss a number of alternative criteria but state that "squared error is analytically convenient and is the most popular" [HA01]. The general problem is to minimize a *loss function* $L(Y, f(X))$ that penalizes prediction errors. The least squares loss function is $\Sigma(Y - f(X))^2$ but other loss functions (e.g. $\Sigma \mid Y - f(X) \mid$) can also be used. A different approach to classification problems is based upon *nearest neighbors*. This approach is considered in Section 7.8.

To understand how one would apply least squares to classification problems, consider the data in Table 2.8.1 and shown in Figure 2.8.1. The **X** vector of the independent variables is two dimensional. The first six points in the table are the learning points used to create the model and the last three points are the test points used to check to see how the model predicts on data not used in the development of the model. The **Y** variable used in the least squares analysis is the *class* of the data points (i.e., 0 or 1).

Point	Type	x_1	x_2	Y = Class
1	Learning	0.50	0.00	0
2	Learning	0.75	0.25	0
3	Learning	1.00	0.50	0
4	Learning	0.00	0.50	1
5	Learning	0.25	0.75	1
6	Learning	0.50	1.00	1
7	Test	0.50	0.25	0
8	Test	0.50	0.75	1
9	Test	0.75	0.50	1

Table 2.8.1   Data for a 2D classification problem

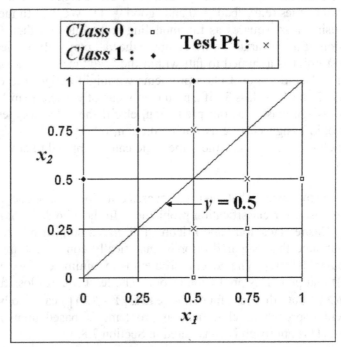

Figure 2.8.1   Display of Data from Table 2.8.1

Assuming a linear model, the following equation is fit to the data:

$$y = a_1 x_1 + a_2 x_2 + a_3 \qquad (2.8.1)$$

Weighting all points equally, the least squares solution is:

$$y = -x_1 + x_2 + 0.5 \qquad (2.8.2)$$

The value of y for test point 7 is -0.5 + 0.25 + 0.5 = 0.25 and since this number is less than 0.5 this point would be classified correctly as belonging to class 0. The value of y for test point 8 is 0.75 so this point would be classified correctly as belonging to class 1. Test point 9 would be misclassified as belonging to Class 0. In Figure 2.8.1 we see the line in the x_1 - x_2 plane in which y is exactly 0.5. Any test point falling above this line would be classified as belonging to class 1 and any point below the line would be classified as belonging to class 2.

Clearly the simple linear model can be extended to d dimensions:

$$y = a_1 x_1 + a_2 x_2 + ... + a_d x_d + a_{d+1}$$ (2.8.3)

Another alternative is to use a higher order model. For example, a 2nd order polynomial model in two dimensions would be:

$$y = a_1 x_1 + a_2 x_2 + a_3 x_1^2 + a_4 x_2^2 + a_5 x_1 x_2 + a_6$$ (2.8.4)

In Figure 2.8.2 Equation 2.8.4 has been fit to 8 data points from class 0 and 10 from class 1. This model predicts that points falling in the shaded area are from Class 0 and otherwise they are from Class 1. Note that 3 of the Class 1 learning points fall within the Class 0 region and one of the Class 0 learning points (the point at 0.75, 0.75) falls within the Class 1 region. Also note that two of the learning points are from different classes although both are located at approximately the same position (near 0.5, 0.5).

The main problem with classification models is the fraction of misclassifications. For example, for a two class problem (Classes 0 and 1) what fraction of Class 0 cases is misclassified as Class 1 and visa-versa? For some problems it is important to reduce misclassifications in one direction and less important in the other direction. For example, consider a problem in which we are building a model to decide which people to inoculate against a certain disease. The two classes are those who have a high probability (Class 1) and a low probability (Class 0) of getting the disease. Clearly, it is much more important to reduce misclassifications for people in Class 1.

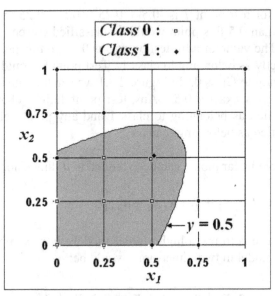

Figure 2.8.2   Fit to 8 Class 0 and 10 Class 1 Points Using Eq. 2.8.4

In Figures 2.8.1 and 2.8.2 the two-dimensional independent variable space was separated into two classes by a single line. For three-class problems two lines would be required. For three-dimensional spaces, two-dimensional surfaces are used to separate the classes. In general for d dimensional spaces, d-1 dimensional surfaces are used to separate the classes. The number of surfaces is **num_classes** -1. If there is only one independent variable x, the separations are just **num_classes** -1 points along the x axis. If there are only two classes then the separation is just a single point. The subject of misclassification can be illustrated using a two-class problem and a single independent variable. Fitting a straight line to the data we get a simple equation relating y and x : $y = a_1 + a_2 x$. The value at which y is 0.5 is x_s (the value of x separating the two classes):

$$x_s = (0.5 - a_1) / a_2 \tag{2.8.5}$$

Assume that the Class 1 data points are normally distributed about $x = 1$, the Class 0 data points are normally distributed about $x = -1$ and the values of σ for both distributions are one. The two distributions are seen in Figure 2.8.3. For cases in which the number of data points in each class is n, then as n becomes large, the line fitted to the data approaches $0.5 + 0.25x$ and from Equation 2.8.5, the values of x_s approaches zero. The misclassification rate for both classes should be about 15.9% because the fraction of a normal distribution beyond one σ is 15.9%. From Figure 2.8.3, the fraction of Class 0 points that are correctly classified is the shaded fraction under the Class 0 distribution. The unshaded fraction to the right of $x = 0$ is the misclassification fraction and is equal to about 0.159.

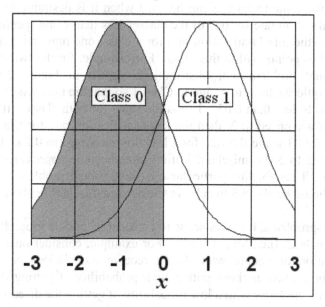

Figure 2.8.3   Values of x for Both Classes are taken randomly from Normal Distributions with $\sigma = 1$

Another important consideration is the relative number of data points in each of the classes. What happens when the x values are taken from the same distributions but there are many more from one class than the other? For example assume that 90% of the data points come from Class 1 (Figure 2.8.3). Running a simulation using 10000 points, the fitted line was $0.792 + 0.134x$ and the value of x_s was -2.183. This value is over one σ to the left of the center of the Class 0 distribution and as a result over 88% of the Class 0 test points were misclassified while only 0.03% of the Class 1 test points were misclassified. These results show that the method described above must be modified if there are a significantly unequal number of learning points in the different classes. There is a simple solution to this problem! By adjusting the weights used for each point based upon the relative class populations, the separation can be accomplished so that the misclassification rates can be approximately equalized. For example, for the above simulation with 90% Class 1 data points, when the weight for Class 0 was increased to 9 (compared to 1 for Class 1), the fitted line was close to the $y = 0.5 + 0.25x$ line noted when the number of data points in each class were equal. Thus the value of x_s was close to zero and the misclassification rate for both classes was about 15.9%.

Clearly, the same technique can be used when it is desirable to achieve a misclassification rate for one of the classes less than some specified value. To reduce the misclassification rate for a class, one must raise the weight attributed to points within this class. For example, for the two-class, one-dimensional problem with distributions shown in Figure 2.8.3, what weight would we have to assign to Class 1 so that the misclassification rate is reduced to less than 0.05? If n is the same for both classes, if the Class 1 weights are raised to 2, then the misclassification rate for Class 1 is reduced to 0.088 while the rate for Class 0 is increased to 0.26. Increasing the weights to 3 the misclassification rates become approximately 0.046 and 0.36. Thus for this particular one dimensional problem the Class 1 weight should be about 3 to achieve the desired misclassification rate.

For some problems, it makes sense to consider a "middle ground": Class 0, Class 1 or N.C. (i.e., "Not Clear"). For example, consider once again the classification of people who should receive inoculation for a disease. Class 1 is defined as those with a high probability of getting the disease and Class 0 are those with a low probability of getting the disease.

δ	Correct	Not Clear	Misclassified
0.00	0.841	0.000	0.159
0.05	0.790	0.095	0.115
0.10	0.729	0.190	0.081
0.15	0.658	0.288	0.054
0.20	0.581	0.382	0.037
0.25	0.498	0.478	0.024

Table 2.8.1   Correct, Not-Clear and Misclassified Rates as a Function of δ for Figure 2.8.3 distributions. Classification is Not-Clear when $0.5 - \delta < y < 0.5 + \delta$.

Let us say that there is a more expense test to decide whether or not a person should be inoculated. This test would only be used for those falling in the N.C. category. The criterion for identifying a person as falling within Class 1 would be a value of $y \geq 0.5 + \delta$, Class 0 for $y \leq 0.5 - \delta$, and N.C. for $0.5 - \delta < y < 0.5 + \delta$. The value of δ would be set based upon the acceptable misclassification rates. For the class distributions considered in

Figure 2.8.3, the effect of δ is seen in Table 2.8.1. We see that for these distributions, a value of δ of 0.25 results in a situation in which the model properly identifies only about half of the people examined. About 2.4% are misclassified and the remaining people fall within the N.C. category.

Another problem encountered when attempting to find a point or line or surface to separate classes is based upon the distribution of the classes within the independent variable space. There are distributions that can't be separated so simply. For example, consider the two-class problem in Figure 2.8.4. There are two independent variables and the two classes are distributed in such a way so that separation of the classes cannot be accomplished with a single line. Problems such as this are better handled using a "nearest neighbor" approach as described in Section 7.8.

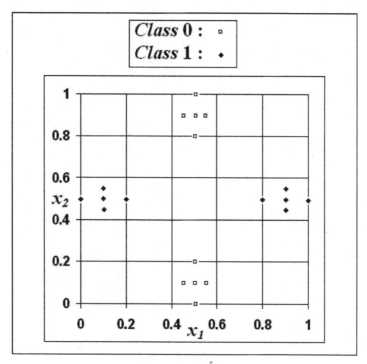

**Figure 2.8.4   Problematic Distribution: Separation cannot be ac-
complished with a single line**

Chapter 3 MODEL EVALUATION

3.1 Introduction

Once a least squares analysis has been completed, we turn our attention to an evaluation of the results. Is the model an adequate representation of the data? Modeling data is not always based upon a "correct" mathematical model. Sometimes one is interested in comparing alternative theoretical models to determine which theory is most applicable to the experimental data. Sometimes the model is proposed as a series and one needs to make a decision regarding the number of terms to keep to best represents the data. There are many situations in which all that one is interested in is an analytical equation that can be used to describe the data. One might start with a simple model and then progressively add terms. At what point do the additional terms lead to a poorer model?

If the data is to be analyzed using the method of least squares, and if we have n data points, the maximum number of unknown parameters that can be determined is $n-1$. If we also have n_b Bayesian estimators, then the maximum is increased to $n+n_b-1$. As the number of unknown parameters is increased, S (the weighted sum of the residuals) decreases so at first glance one might think that the more unknown parameters included in the model, the better the fit. However, we reach a point where additional terms begin to model the noise in the data rather than the true signal. Fortunately, statistical methods are available for determining when we should stop adding terms to a model. In this chapter, statistical methods for evaluation of models are presented.

3.2 Goodness-of-Fit

In Section 1.3 the χ^2 (chi-squared) distribution was discussed. Under certain conditions, this distribution can be used to measure the **goodness-of-fit** of a least squares model. To apply the χ^2 distribution to the measurement of goodness-of-fit, one needs estimates of the uncertainties associated with the data points. In Sections 2.5 and 2.6 it was emphasized that only relative uncertainties were required to determine estimates of the uncertainties associated with the model parameters and the model predictions. However, for goodness-of-fit calculations, **estimates of absolute uncertainties are required.** When such estimates of the absolute uncertainties are unavailable, the best approach to testing whether or not the model is a good fit is to examine the residuals. This subject is considered in Section 3.9.

The goodness-of-fit test is based upon the value of $S/(n\text{-}p)$. Assuming that S is based upon reasonable estimates of the uncertainties associated with the data points, if the value of $S/(n\text{-}p)$ is much less than one, this usually implies some sort of misunderstanding of the experiment. If the value is much larger than one, then one of the following is probably true:

1) The model does not adequately represent the data.

2) Some or all of the data points are in error.

3) The estimated uncertainties in the data are erroneous.

Assuming that the model, the data and the uncertainty estimates are correct, the value of S (the weighted sum of the residuals) will be distributed according to a χ^2 distribution with $n\text{-}p$ degrees of freedom. Since the expected value of a χ^2 distribution with $n\text{-}p$ degrees of freedom is $n\text{-}p$, the expected value of $S/(n\text{-}p)$ is one. If one assumes that the model, data and estimated uncertainties are correct, the computed value of $S/(n\text{-}p)$ can be compared with $\chi^2/(n\text{-}p)$ to determine a probability of obtaining or exceeding the computed value of S. If this probability is too small, then the goodness-of-fit test fails and one must reconsider the model and/or the data.

Counting experiments in which the data points are numbers of events recorded in a particular time window are a class of experiments in which estimates of the absolute uncertainty associated with each data point are available. Let us use Y_i to represent the number of counts recorded in the time interval t_i. According to Poisson statistics the expected value of σ_i^2 (the variance associated with Y_i) is just Y_i and the weight associated with this point is $1/Y_i$. From Equation 2.2.1 we get the following expression for S:

$$S = \sum_{i=1}^{i=n} w_i R_i^2 = \sum_{i=1}^{i=n} w_i (Y_i - y_i)^2 = \sum_{i=1}^{i=n} (Y_i - f(t_i))^2 / Y_i \qquad (3.2.1)$$

Since the expected value of $(Y_i - y_i)^2$ is $\sigma_i^2 = Y_i$ the expected value of $w_i R_i^2$ is one. The expected value of S is not n as one might expect from this equation. If the function f includes p unknown parameters, then the number of degrees of freedom must be reduced by p and therefore the expected value of S is $n-p$. This might be a confusing concept but the need for reducing the expected value can best be explained with the aid of a qualitative argument. Lets assume that n is 3 and we use a 3 parameter model to fit the data. We would expect the model to go thru all 3 points and therefore, the value of S would be zero which is equal to $n-p$.

To illustrate this process, let us use data included in Bevington and Robinson's book *Data Reduction and Error Analysis* [BE03]. The data is presented graphically in Figure 3.2.1 and in tabular form in Table 3.2.1. The data is from a counting experiment in which a Geiger counter was used to detect counts from an irradiated silver piece recorded in 15 second intervals. The 59 data points shown in the table include two input columns (t_i and Y_i) and one output column that is the residual divided by the standard deviation of Y_i (i.e., R_i / σ_i):

$$R_i / \sigma_i = (Y_i - y_i) / \sqrt{Y_i} \qquad (3.2.2)$$

The data was modeled using a 5 parameter equation that included a background term and two decaying exponential terms:

$$y = a_1 + a_2 e^{-t/a_4} + a_3 e^{-t/a_5} \qquad (3.2.3)$$

i	t_i	Y_i	R_i / σ_i	i	t_i	Y_i	R_i / σ_i
1	15	775	0.9835	31	465	24	-0.0208
2	30	479	-1.8802	32	480	30	1.2530
3	45	380	0.4612	33	495	26	0.7375
4	60	302	1.6932	34	510	28	1.2466
5	75	185	-1.6186	35	525	21	0.0818
6	90	157	-0.4636	36	540	18	-0.4481
7	105	137	0.3834	37	555	20	0.1729
8	120	119	0.7044	38	570	27	1.6168
9	135	110	1.3249	39	585	17	-0.2461
10	150	89	0.4388	40	600	17	-0.1141
11	165	74	-0.2645	41	615	14	-0.7922
12	180	61	-1.0882	42	630	17	0.1231
13	195	66	0.2494	43	645	24	1.6220
14	210	68	1.0506	44	660	11	-1.4005
15	225	48	-1.0603	45	675	22	1.4360
16	240	54	0.2938	46	690	17	0.5068
17	255	51	0.3200	47	705	12	-0.7450
18	270	46	0.0160	48	720	10	-1.3515
19	285	55	1.5750	49	735	13	-0.0274
20	300	29	-2.2213	50	750	16	0.5695
21	315	28	-2.0393	51	765	9	-1.4914
22	330	37	0.0353	52	780	9	-1.4146
23	345	49	2.0104	53	795	14	0.2595
24	360	26	-1.4128	54	810	21	1.7830
25	375	35	0.5740	55	825	17	1.0567
26	390	29	-0.2074	56	840	13	0.1470
27	405	31	0.4069	57	855	12	-0.0891
28	420	24	-0.7039	58	870	18	1.3768
29	435	25	-0.2503	59	885	10	-0.6384
30	450	35	1.6670				

Table 3.2.1   Input data (t and Y) from Table 8.1, Bevington and Robinson [BE03].

The input data in the table was analyzed using the REGRESS program [see Section 6.8] and yielded the following equation:

$$y = 10.134 + 957.77e^{-t/34.244} + 128.29e^{-t/209.68}$$

For example, for the first data point (i.e., $t = 15$), the computed value of y according to this equation is 747.62. The **relative errors** included in the table (i.e., R_i/σ_i) are computed using Equation 3.2.2. Thus the value of the relative error for the first point is $(775 - 747.62)/\sqrt{775} = 0.9835$. Note that the relative errors are distributed about zero and range from -2.2213 to 2.0104. The value of S is the sum of the squares of the relative errors and is 66.08. The number of points n is 59 and the number of unknown parameters p is 5 so the value of $S/(n-p)$ is 1.224. The goodness-of-fit test considers the closeness of this number to the most probable value of one for a correct model. So the question that must be answered is: how close to one is 1.224?

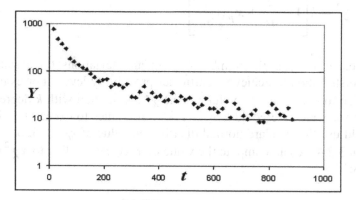

Figure 3.2.1   Bevington and Robinson data [BE03]

In Section 1.3 it was mentioned that for the χ^2 distribution with k degrees of freedom, the mean is k and the standard deviation is $\sqrt{2k}$. Furthermore as k becomes larger, the χ^2 distribution approaches a normal distribution. We can use these properties of the distribution to estimate the probability of obtaining a value of S greater or equal to 66.08 for a χ^2 distribution with 54 degrees of freedom:

$$66.08 = 54 + x_p * \sqrt{2 * 54} = 54 + x_p * 10.39$$

$$x_p = (66.08 - 54)/10.39 = 1.163$$

In other words, 66.08 is approximately $x_p = 1.16$ standard deviations above the expected value of 54. From a table of the normal distribution we can verify that the integral from 0 to 1.163 standard deviations is 0.3777, so the probability of exceeding this value is $0.5 - 0.3777 = 0.1223$ (i.e., about 12%). Typically one sets a value of the probability at which one would reject the model. For example, if this probability is set at 1%, then the lower limit of x_p for rejecting the model would be 2.326. Since our computed value of x_p is much less than 2.326, we have no reason to reject the five parameter model.

If we really want to be pedantic, we can make a more accurate calculation. From the *Handbook of Mathematical Functions* [AB64], Equation 26.4.17 is suggested for values of $k > 30$:

$$\chi^2 = k\left[1 - \frac{2}{9k} + x_p\sqrt{\frac{2}{9k}}\,\right]^3$$

(3.2.4)

In this equation x_p is the number of standard deviations for a standard normal distribution to achieve a particular probability level. For example, if we wish to determine the value of the χ^2 distribution with k degrees of freedom for which we could expect 1% of all values to exceed this level, we would use the standard normal distribution value of $x_p = 2.326$. Using Equation 3.2.4 we can compute the value of x_p corresponding to a χ^2 value of 66.08:

$$66.08 = 54[1 - 2/(9*54) + x_p\sqrt{2/(9*54)}\,]^3$$

Solving this equation for x_p we get a value of 1.148 which is close to the value of 1.163 obtained above using the simple normal distribution approximation.

Equation 3.2.3 is a 5 parameter model that includes a background term and two decaying exponential terms. If we were to simplify the equation to include only a single exponential term, would we still have a "reasonable" model? Would this equation pass the goodness-of-fit test?   The proposed alternative model is:

$$y = a_1 + a_2 e^{-t/a_3}$$

(3.2.5)

Once again using the REGRESS program the resulting equation is:

$$y = 18.308 + 752.99e^{-t/62.989}$$

and the value of S is 226.7. The value of x_p corresponding to this value of S is estimated as follows:

$$226.7 = 54 + x_p * \sqrt{2 * 54}$$

$$x_p = (226.7 - 54) / 10.39 = 16.62$$

This value is so large that we would immediately reject the proposed model. The probability of getting a value of S that is over 16 standard deviations above the expected value for a correct model is infinitesimal.

3.3 Selecting the Best Model

When modeling data, we are often confronted with the task of choosing the best model out of several proposed alternatives. Clearly we need a definition of the word "best" and criteria for making the selection. At first glance one might consider using S (the weighted sum of the squares of the residuals) as the criterion for choosing the best model but this choice is flawed. As p (the number of unknown parameters included in the model) increases, the values of S decreases and becomes zero if p is equal to n (the number of data points).

To illustrate this point, consider the data shown in Figure 2.3.2. This data was generated based upon a parabolic model ($p = 3$) and included 5% random noise. The 10 data points were fit using the following polynomial model with values of p varying from 2 to 8:

$$y = a_1 + \sum_{k=2}^{k=p} a_k x^{k-1} \tag{3.3.1}$$

Results are included in Table 3.3.1. Note that the value of S decreases as p increases but that the minimum value of $S / (n-p)$ is achieved for $p=3$. This is encouraging because the minimum value of $S / (n-p)$ was obtained for the value of p upon which the data was generated. However, can we use this criterion (i.e., choose the model for which $S / (n-p)$ is minimized) as the sole criterion for selecting a model?

p	S	S/(n-p)	RMS-Error	RMS-Rel_Error
2	3619.19	452.418	19.239	19.025
3	6.22	0.888	0.831	0.789
4	6.21	1.035	0.835	0.788
5	5.86	1.171	0.696	0.765
6	5.50	1.375	0.531	0.742
7	5.12	1.706	0.446	0.715
8	3.66	1.829	0.652	0.605

Table 3.3.1   Using Equation 3.3.1 to model data in Table 2.3.1

In this table the **RMS-Error** (the root-mean-square error) is computed as follows:

$$RMS\text{ - }Error = \sqrt{\frac{\sum_{i=1}^{i=n}(Y_i - y_i)^2}{n}} \tag{3.3.2}$$

The **RMS-Rel-Error** (the root-mean-square relative error) is computed as follows:

$$(RMS - Rel - Error)^2 = \frac{\sum_{i=1}^{i=n}((Y_i - y_i)/\sigma_{y_i})^2}{n}$$

Comparing these equations with the definition of S (Equations 2.3.1 and 2.3.3), we see that **RMS-Error** and **RMS-Rel-Error** are just modified forms of S:

$$RMS - Error = \sqrt{S/n} \qquad (w_i = 1) \tag{3.3.3}$$

$$RMS - Rel - Error = \sqrt{S/n} \qquad (w_i = 1/\sigma_{y_i}^2) \tag{3.3.4}$$

Note that for $p=2$ (i.e., a straight line fit to the parabolic data) all the results are terrible. When the model under-fits the data we expect to see large values of $S/(n-p)$ and the **RMS** error measures. However, when the data is over-fitted (i.e., $p > 3$ for this example), S and **RMS-Rel-Error** should decrease with increasing p. (Since S was based upon statistical weighting, we expect **RMS-Rel-Error** to decrease monotonically with increasing p

but **RMS-Error** does not necessarily decrease monotonically.) In Figure 3.3.1 the data is shown with the curves for **p** = 3 and **p** = 8. Note that the two curves appear quite similar up to about **x** = 8 but this is an illusion due to the scale of the graph. In reality for **p** = 8, the fitted curve is actually modeling the noise in the data. We see this quite clearly for **x** greater than 8.

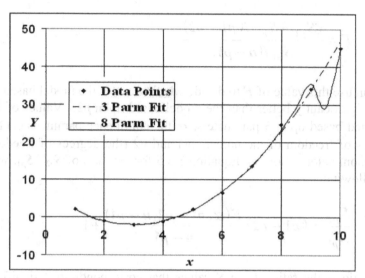

Figure 3.3.1   Table 3.3.1 Least Squares curves for p = 3 and 8

p	S	S/(n-p)	RMS-Error	RMS-Rel Error
2	359971.1	400.4131	18.1951	19.9881
3	873.8	0.9730	0.9368	0.9848
4	873.0	0.9732	0.9306	0.9843
5	871.3	0.9724	0.9252	0.9834
6	871.2	0.9734	0.9247	0.9833
7	871.0	0.9743	0.9249	0.9832
8	870.9	0.9753	0.9247	0.9831

Table 3.3.2   Using Equation 3.3.1 to model parabolic data with 5% random noise and n=901 ranging from 1 to 10 with increments of 0.01.

We have established that S is distributed according to a χ^2 distribution with $n-p$ degrees of freedom. Let us consider a series of models in which an increase in p represents an increase in the complexity of the model. In other words, if we increase p from 3 to 4, we are adding an additional term to the 3 parameter model. Let us use the notation S_{p1} and S_{p2} to represent two values of S with differing values of p but based upon the same n data points. Choosing $p1$ to be greater than $p2$, then $S_{p2} > S_{p1}$. An interesting and useful property of χ^2 distributions is the following: $S_{p2} - S_{p1}$ is distributed according to a χ^2 distribution with $p1$-$p2$ degrees of freedom [FR92]. In Section 1.3 the F distribution was defined as the ratio of two χ^2 distributions divided by their degrees of freedom. Thus the following ratio should follow an F distribution:

$$F = \frac{(S_{p2} - S_{p1})/(p1 - p2)}{S_{p1}/(n - p1)} \qquad (3.3.5)$$

We can use this value of F to decide whether or not the model based upon $p1$ is significantly better than the model based upon $p2$. Values of F are tabulated based upon 3 parameters, α (the confidence parameter), $v1$ (the degrees of freedom of the numerator) and $v2$ (the degrees of freedom of the denominator). Solving Equation 3.3.5 for the ratio of S_{p2} / S_{p1}, we get the following:

$$\frac{S_{p2}}{S_{p1}} = (p1 - p2)\frac{F(\alpha, p1 - p2, n - p1)}{n - p1} + 1 \qquad (3.3.6)$$

This ratio is the ratio of the S values that corresponds to a significance level of α. In other words, if the extra $p1$-$p2$ terms in the $p1$ parameter model neither add nor detract from the original $p2$ model, we would expect the ratio to exceed this value $100\alpha\%$ of the time. To illustrate how this Equation is used, let us first compare the two values of S for $p1 = 3$ and $p2 = 2$ from Table 3.3.2. Let us use a value of $\alpha = 0.01$. Table 3.3.2 is based upon a value of n - $p1 = 898$ which is very large and we can therefore use the following approximation:

$$F(\alpha, v_1, v_2) \rightarrow \chi^2(\alpha, v_1)/v_1 \quad \text{as} \quad v_2 \rightarrow \infty \qquad (3.3.7)$$

The value of $F(0.01, 1, 898)$ is thus approximately $\chi^2(0.01, 1) / 1 = 6.63$ so we would expect that if the model with $p1 = 3$ parameters is significantly better than the model with two parameters, then the ratio S_{p2} / S_{p1} would be greater than :

$$\frac{S_2}{S_3} = (3 - 2)\frac{6.63}{901 - 3} + 1 = 1.007$$

The observed ratio (i.e., 359971 / 873.8) is about 412 which is much greater than 1.007 so the 3 parameter model is clearly an improvement upon the two parameter model. We can repeat the analysis comparing the 4 and 5 parameter models to the 3 parameter models:

$$\frac{S_3}{S_4} = (4-3)\frac{6.63}{901-4} + 1 = 1.007 \qquad \text{(1\% confidence limit)}$$

$$\frac{S_3}{S_5} = (5-3)\frac{4.61}{901-5} + 1 = 1.010 \qquad \text{(1\% confidence limit)}$$

(The value of $F(0.01, 2, 896)$ is approximately 4.61.) The measured values of these ratios are 873.8 / 873.0 = 1.001 and 873.8 / 871.3 = 1.003. Since neither of these values exceeds the relevant 1% confidence limit we conclude that the reduction in S going to 4 and 5 parameter models is not significant. In other words, increasing the number of unknown parameters beyond 3 does not yield a significantly better model.

We can also use the previous test to consider totally different models with differing values of $p1$ and $p2$. For such cases, if we choose $p1$ to be greater than $p2$, S_{p2} is not necessarily greater than S_{p1}. Clearly, if $S_{p2} \le S_{p1}$ then we would immediately choose Model 2 as not only does it have less parameters, it also exhibits a smaller value of S.

Another question that we should consider is how do we compare models in which the numbers of unknown parameters are the same? Clearly, since the degrees of freedom are the same for both models, a direct comparison of the S values is sufficient to determine which is the better model. However, when the values of S are close, we should then consider the issue of significance. Let us define Model 1 as the model with the smaller value of S. We can then ask the question: is Model 1 significantly better than Model 2? Once again, our test of significance is based upon the F distribution. Let us use the notation S_1 and S_2 to represent the weighted sum of the squares for the two models, both with $n - p$ degrees of freedom. Regardless of whether or not the weights are based upon actual or relative σ's, the ratio S_2 / S_1 is F distributed:

$$F = \frac{S_2}{S_1} \qquad\qquad (3.3.8)$$

Our significance test for this case is simply:

$$\frac{S_2}{S_1} = F(\alpha, v, v) \qquad\qquad (3.3.9)$$

where v is the number of degrees of freedom (i.e., $n - p$). Values of $F(\alpha, v, v)$ are included in Table 3.3.3 for various values of v and for values of α equal to 0.01 and 0.05.

To illustrate the use of Table 3.3.3, consider the 63 data points shown in Figure 3.3.2. We have no information regarding the uncertainty associated with these data points so the analysis is based upon unit weighting. In Table 3.3.4 the values of $S/(n-p)$ are listed for 3 different models each with $n-p = 60$.

v	$F(0.05, v, v)$	$F(0.01, v, v)$
5	5.05	11.00
10	2.98	4.85
15	2.40	3.52
20	2.12	2.94
30	1.84	2.39
40	1.69	2.11
60	1.53	1.84
120	1.35	1.53

Table 3.3.3   Values of $F(\alpha, v, v)$

Model Number	Model	$S/(n-p)$
1	$a_1 \exp(-a_2 x) + a_3$	33567
2	$a_1 x + a_2 x^2 + a_3$	86385
3	$a_1/x + a_2/x^2 + a_3$	1170356

Table 3.3.4 $S/(n-p)$ values for 3 models for Figure 3.3.2 data

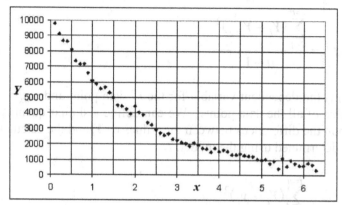

Figure 3.3.2   Input Data for Models in Table 3.3.4

Since unit weighting was used, the absolute values of $S / (n\text{-}p)$ are not meaningful. Clearly Model 3 is much worse than either Model 1 or Model 2. We can however use Equation 3.3.9 to decide whether or not Model 1 is significantly better than Model 2. The value of S_2 / S_1 is 2.57 and the number of degrees of freedom $(n\text{-}p)$ is 60. From Table 3.3.3 we see that 2.57 is far above the confidence limit of 1.84 for $\alpha = 1\%$, so we can conclude that Model 1 is significantly better than Model 2. Since the absolute value of $S / (n\text{-}p)$ is meaningless, we can't say that Model 1 is the "correct" model for this data. All we can say is that it is significantly better than the other 2 models considered.

3.4 Variance Reduction

Variance reduction (**VR**) is one of the most commonly used measures of the value of a model. The most attractive feature of **VR** is that it can be used for any model, linear or nonlinear, in which one or more independent variables are used to describe the behavior of a dependent variable **Y**. **VR** is typically defined as the percentage of the variance in the dependent variable that is explained by the model. Variance is defined as the standard deviation squared, so the variance of the dependent variable is computed as follows:

$$\sigma^2 = \frac{\sum_{i=1}^{i=n}(Y_i - Y_{avg})^2}{n-1} \tag{3.4.1}$$

where Y_i is the dependent variable for the ith data point and Y_{avg} is the average value of all the data points. We can define a similar quantity that is based upon the differences between the values of Y_i and the calculated values y_i as determined using the model:

$$\sigma_m^2 = \frac{\sum_{i=1}^{i=n}(Y_i - y_i)^2}{n-1} \tag{3.4.2}$$

Let us call this quantity the **model variance**. Variance reduction is based upon the ratio of the model variance to the variance in the dependent variable and is computed as follows:

$$VR = 100 * \left(1 - \frac{\sigma_m^2}{\sigma^2}\right) = 100 * \left(1 - \frac{\sum_{i=1}^{i=n}(Y_i - y_i)^2}{\sum_{i=1}^{i=n}(Y_i - Y_{avg})^2}\right) \tag{3.4.3}$$

From this equation we see that if we have a **perfect model** (i.e., the calculated values of y_i are exactly equal to the actual values of Y_i) then the value of VR is exactly equal to 100. If the model is useless (i.e., has no predictive power), then we would expect a value of VR approximately equal to zero. Models can actually be negative, implying that the values predicted by the model are worse than just using Y_{avg} as the predictor for all points.

As an example, consider the data in Table 2.3.1. Table 3.4.1 summarizes the values of VR obtained for two different models and two different weighting schemes:

Model	$w_i = 1$	$w_i = 1/\sigma^2_{y_i}$
$y = a_1 + a_2x$	80.29	-48.19
$y = a_1 + a_2x + a_3x^2$	99.76	99.72

Table 3.4.1   Values of *VR* for data in Table 2.3.1

For the 3 parameter models the results in Table 3.4.1 show that both weighting schemes explain more than 99% of the variance in the data. These high values of *VR* are reasonable because the data was based upon a parabolic model with some random noise and regardless of the weighting scheme, both models reasonably represent the data. However, the results for the 2 parameter models are vastly different. The model obtained using unit weighting explains more than 80% of the variance but the model obtained using statistical weighting is terrible. To understand why this happened, note that the value of *VR* is maximized when the model variance is minimized. Examining Equation 3.4.2 it can be seen that if unit weighting (i.e., $w_i = 1$) is used, the model variance is just $S/(n-1)$. Thus the model obtained using unit weighting is also the model that minimizes model variance. Does this imply that unit weighting is preferable to statistical weighting? Not at all! It only implies that *VR* is a useful measure of model performance only when all data points are equally (or approximately equaled) weighted.

It is useful to be able to apply a test of significance to values of *VR*. The usual procedure for testing the significance of a model is to test the **null hypothesis**: if the model has no predictive value, what is the value of *VR* that we would exceed 100α % of the time? The model variance is χ^2 distributed with n-p degrees of freedom and the variance in the data is χ^2 distributed with n-1 degrees of freedom, therefore the following ratio is *F* distributed:

$$F = \frac{\sum_{i=1}^{i=n}(Y_i - Y_{avg})^2/(n-1)}{\sum_{i=1}^{i=n}(Y_i - y_i)^2/(n-p)} \qquad (3.4.4)$$

Solving for the ratio of the variances:

$$\frac{\sum\limits_{i=1}^{i=n} (Y_i - y_{avg})^2}{\sum\limits_{i=1}^{i=n} (Y_i - y_i)^2} = \frac{n-p}{n-1} F(\alpha, n-1, n-p) \tag{3.4.5}$$

For models with some value, this ratio will be greater than one. If the ratio is less than one, then the model exhibits negative **VR** and can be considered worthless. We can use this equation to test the significance of the 2 parameter model based upon unit weighting.   Let us choose a value of $\alpha = 0.01$. The value of $F(0.01, 9, 8) = 5.91$ can be determined from appropriate tables [AB64,FR92].   Thus we would expect that the variance ratio would exceed 8*5.91 / 9 = 5.25 one percent of the time if the null hypothesis is true (i.e., the model has no predictive power). The value of **VR** corresponding to this ratio is 100 * (1 – 1/5.25) = 80.9. Surprisingly, the value of **VR** obtained for the 2 parameter model with unit weighting does not pass this significance test although it is very close (i.e., 80.29). When the number of degrees of freedom for the model is small, it is possible to get large values of **VR** even for useless models. However, as the number of data points increases, the upper limit for the null hypothesis decreases dramatically. For example, if the value of **VR** = 80.29 was based upon 30 points, then $F(0.01, 29, 28) = 2.40$ which corresponds to a ratio of 2.32 (from Equation 3.4.5) which corresponds to a **VR** = 100 * (1 – 1/2.32) = 56.8. A value of 80.29 would therefore be highly significant.

3.5 Linear Correlation

Linear correlation is a concept used to measure the linear (straight-line) relationship between variables. The **correlation coefficient** ρ can vary between -1 and 1. If $\rho = 0$ the variables are unrelated (in the linear sense). If $\rho = 1$ the variables fall exactly upon a straight line with a positive slope and if $\rho = -1$ they fall upon a straight line with a negative slope. Typical data for several different values of ρ are seen in Section 2.5 (Figure 2.5.1). If we have two variables u and v, the correlation coefficient is defined as follows:

$$\rho = \frac{\sigma_{uv}}{\sigma_u \sigma_v} \tag{3.5.1}$$

The terms σ_u and σ_v are standard deviations of the variables u and v and σ_{uv} is called the **covariance** between u and v. The covariance is defined as follows:

$$\sigma_{uv} = \int_{-\infty}^{\infty} \int_{-\infty}^{\infty} (u - \mu_u)(v - \mu_v) \Phi(u,v) du dv \qquad (3.5.2)$$

where $\Phi(u,v)$ is the **bivariate** distribution function for the variables u and v, and μ_u and μ_v are the mean values of u and v. When the covariance is close to zero, then a straight line relationship between u and v is not a reasonable assumption. If we have paired data $\{u_i, v_i ; i = 1 .. n\}$, we can compute r (an unbiased estimate of ρ) as follows:

$$r = \frac{\displaystyle\sum_{i=1}^{i=n} (u_i - u_{avg})(v_i - v_{avg})}{\sqrt{\displaystyle\sum_{i=1}^{i=n}(u_i - u_{avg})^2} \sqrt{\displaystyle\sum_{i=1}^{i=n}(v_i - v_{avg})^2}} \qquad (3.5.3)$$

To facilitate the calculation we use the following equalities:

$$\sum_{i=1}^{i=n} (u_i - u_{avg})(v_i - v_{avg}) = \sum_{i=1}^{i=n} u_i v_i - n u_{avg} v_{avg} \qquad (3.5.4)$$

$$\sum_{i=1}^{i=n} (u_i - u_{avg})^2 = \sum_{i=1}^{i=n} u_i^2 - n u_{avg} \qquad (3.5.5)$$

$$\sum_{i=1}^{i=n} (v_i - v_{avg})^2 = \sum_{i=1}^{i=n} v_i^2 - n v_{avg} \qquad (3.5.6)$$

As an application of the use of Equation 3.5.3, consider the data in Table 3.5.1. This data includes the heights (in meters) and weights (in kilograms) of 7 individuals on a basketball team.

i	u = height	u*u	v = weight	v*v	u * v
1	1.87	3.4969	83	6889	155.21
2	1.92	3.6864	97	9409	186.24
3	2.04	4.1616	86	7396	175.44
4	2.10	4.4100	105	11025	220.50
5	1.98	3.9204	101	10201	199.98
6	2.02	4.0804	92	8464	185.84
7	1.77	3.1329	79	6241	139.83
Sum	13.70	26.8886	643	59625	1263.04
Avg	1.9571	3.8412	91.857	8517.9	180.434

Table 3.5.1   Heights and Weights for 7 Basketball Players.

Using Equation 3.5.3 thru 3.5.6 the correlation coefficient is:

$$r = \frac{1263.04 - 7 * 1.9571 * 91.857}{\sqrt{26.8886 - 7 * 19571^2} \sqrt{59625 - 7 * 91.857^2}} = 0.705$$

It can be shown that r^2 is the same as variance reduction **VR** (expressed as a fraction) when the equation for y is a straight line using unit weighting [FR92]. In other words, r^2 is the fraction of the variance in the data explained by the unit weighted least squares straight line. Applying least squares to the data in Table 3.5.1 we get the following linear relationship between **height** and **weight**:

$$height = 1.20422 + 0.0081966 * weight$$

Using this equation, the **VR** computed using Equation 3.4.3 is 49.7 which expressed as a fraction is 0.497 = 0.705 * 0.705. Thus we see that r = 0.705 implies that approximately 50% of the variance in the data is accounted for by this line.

The correlation coefficient is a measure of the linear relationship between two variables. It does not answer the question: are the two variables related? Consider, for example, the data in Table 3.5.2.

i	y	y*y	x	x*x	x*y
1	19	384	-3	9	-57
2	9	81	-2	4	-18
3	3	9	-1	1	-3
4	1	1	0	0	0
5	3	9	1	1	3
6	9	81	2	4	18
7	19	384	3	9	57
Sum	63	949	0	28	0
Avg	9	135.57	0	4	0

Table 3.5.2 $y = 2x^2 + 1$ (a perfect parabolic fit)

Note the perfect parabolic relationship between the dependent variable y and the independent variable x. Using y as the u variable and x as the v variable, and substituting into Equations 3.5.3 through 3.5.6, we compute r as follows:

$$r = \frac{0 - 7 * 9 * 0}{\sqrt{949 - 7 * 9^2} \sqrt{28 - 7 * 0^2}} = 0.0$$

Although the data exhibits a perfect relationship between x and y, we compute a correlation coefficient of zero. If however, we use the relationship $y = a_1 + a_2x + a_3x^2$ and apply least squares to the data in Table 3.5.1, we obtain the values $a_1 = 1$, $a_2 = 0$ and $a_3 = 2$ with $VR = 100$. We can use correlation to compare the actual values of y with the computed values. Since the relationship is a perfect fit, $Y_i = y_i$ for all points. The data is summarized in Table 3.5.3:

i	Y (actual)	Y*Y	y (computed)	y*y	Y*y
1	19	384	19	384	384
2	9	81	9	81	81
3	3	9	3	9	9
4	1	1	1	1	1
5	3	9	3	9	9
6	9	81	9	81	81
7	19	384	19	384	384
Sum	63	949	63	949	949
Avg	9	135.57	9	135.57	135.57

Table 3.5.3   Comparing actual and computed values of y

The computed value of r is one:

$$r = \frac{949 - 7 * 9 * 9}{\sqrt{949 - 7 * 9^2} \sqrt{949 - 7 * 9^2}} = 1.0$$

Thus for relationships that are other than straight lines, correlation is meaningful only when comparing actual and computed values of the variable.

We next turn our attention to the significance of the computed value of the correlation coefficient. We first define a new variable z based upon r:

$$z = \frac{1}{2} ln(\frac{1+r}{1-r}) \tag{3.5.7}$$

Remembering that ρ is the true value of the correlation coefficient, it can be shown that z is approximately normally distributed with the following properties [WI62, FR92]:

$$\mu_z = \frac{1}{2} ln(\frac{1+\rho}{1-\rho}) \tag{3.5.8}$$

$$\sigma_z^2 = \frac{1}{n-3} \tag{3.5.9}$$

As an example, we can apply these equations to the data included in Table 3.5.1. The null hypothesis for this example is that the heights and weights of the basketball players are uncorrelated (i.e., ρ is zero). If this hypothesis is true, the mean of the distribution (from 3.5.8) is zero and the variance (from 3.5.9) is 1/4. The value of r is 0.705, so from Equation 3.5.7 the value of z is 0.5 * $ln(1.705/0.295) = 0.877$. The following parameter should be t distributed with n - 3 degrees of freedom:

$$t = \frac{z(r) - z(\rho)}{\sigma_z} = \frac{0.877 - 0}{\sqrt{1/(7-3)}} = \frac{0.877}{0.5} = 1.75$$

At a 5% level of confidence the value of t would have to exceed 2.132 to be deemed significant, so there is a probability greater than 5% that the computed value of $r = 0.705$ could have happened by chance. If the computation had been based upon 4 times as many data points (i.e., $n = 28$) then the t parameter would be $0.877/0.2 = 4.385$ which is highly significant even at a confidence level of 0.5%.

In summary, when one is interested in building a model to compute y as a function of an independent variable x, the linear correlation coefficient is only useful when the proposed model is a straight line. The linear correlation coefficient is a measure of how close the relationship is to a straight line and the sign indicates the slope of the line. However, if the model is something other than a straight line, linear correlation is not a useful measure of the power of the model. When a model other than a straight line is proposed, the linear correlation coefficient can be used to compare actual and computed values of the dependent variable y.

3.6 Outliers

The term **outlier** is used to denote a data point that differs considerably from the bulk of the data. We see an example of an outlier in the value of Y_5 in Table 2.3.3. This is an example of an outlier in the dependent variable. Of course, there can also be outliers in the independent variable or variables. An outlier may be due to an error in data collection or it might represent a true event. If an outlier is identified and if it is determined to be erroneous, then the situation can usually be rectified by either correcting the error or rejecting the data point. However, when an outlier represents a true event, we might prefer not to reject the data point. Unfortunately, if we are using least squares analysis of the data, the outlier might cause results that are highly skewed in the direction that accommodates the outlier.

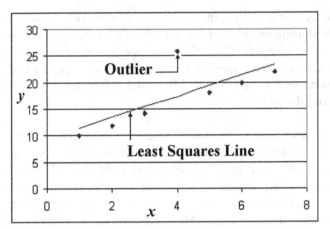

Figure 3.6.1   Data with a Single Outlier

To illustrate this point, consider the data shown in Figure 3.6.1 and tabulated in Table 3.6.1. Notice in the figure that the signs of the residuals (i.e., $Y_i - y_i$) for all points other than the outlier have the same sign.

x	Y (*actual*)	y (*computed*)	y (*exclude x=4*)
1	10.0	11.41	9.99
2	11.8	13.41	11.99
3	14.2	15.40	13.99
4	25.9	17.40	excluded
5	18.1	19.40	17.98
6	19.8	21.39	19.98
7	22.0	23.39	21.97

Table 3.6.1   Figure 3.6.1 data.   Point 4 is an outlier.

Fitting this data with a straight line and weighting all points equally, the resulting line is:

$$y = 9.4143 + 1.9964x$$

and the *VR* (variance reduction) is 56.9%. Eliminating point 4 and repeating the analysis, the resulting line is:

$$y = 7.9976 + 1.9964x$$

and the *VR* is 99.9%!!

As an example of an application that includes a significant number of out-liers, consider the problem of modeling one day fractional changes in the prices of shares appearing on the major stock exchanges. If our database includes several thousands of companies listed on the exchanges, there are usually several everyday that exhibit extremely large changes (either posi-tive or negative) based upon some sort of news or announcement. For ex-ample, when the Enron scandal first made the news, the price of the Enron shares plunged. Positive news, (for example FDA approval for a new drug) can also cause huge increases in the price of the shares of the com-pany mentioned in the announcement. To reduce the effect of these outly-ing events, one strategy is to **clip** the data. Clipping implies setting upper and lower limits on the changes included in the data to be modeled. For example, for stock market modeling, a typical strategy is to clip all points that exhibit a fractional increase greater than 0.25 to 0.25 and all points that exhibit a fractional decrease greater than -0.25 to -0.25. As an exam-ple, if Company X is granted a patent and the stock rises 37% when the news is announced, the fractional change used in the analysis for that date would be reduced to 0.25 instead of the change of 0.37 actually observed.

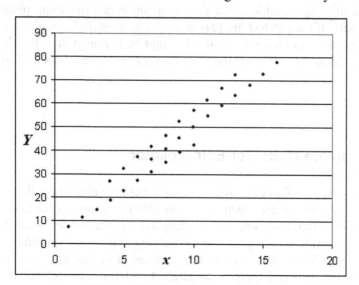

Figure 3.6.2   Clustered Data

Figure 3.6.2 illustrates a different problem associated with outliers. In this figure the data seems to include 3 clusters of data each falling upon a dif-ferent straight line. Some possible explanations come to mind when exam-ining such data:

1) The differences might have been due to the use of several different measuring instruments exhibiting differing biases.

2) The measurements were taken at different times or locations and were affected by changes in a variable not recorded in the data set. For example, the measurements might be temperature sensitive and the explanation for the differences might be simply changes in temperature.

3) The clusters might be real and due to some other overlooked variable. Sometimes strange results such as seen in Figure 3.6.2 lead to interesting discoveries. However, the usual explanation is that there is some problem with the experiment.

If the first explanation explains the clustering, then the experimenter must pay more attention to the calibration of the instruments and perhaps repeat the experiment or correct the data. If the second explanation explains the clustering, then one must determine the cause of the differences and include the missing variable in the data acquisition and modeling phases of the analysis. If the clusters are real and are due to excluded regions in one or more of the variables, one approach might be to remap the variable values. Pyle discusses different strategies for remapping variables [PY99].

3.7 Using the Model for Extrapolation

Often the purpose of an experiment is to predict the value of the dependent variable at a value of the independent variable that cannot be obtained experimentally. Extrapolation can be dangerous if the model is not based upon a true understanding of the underlying theoretical considerations. For some problems where one is only interested in finding a model that represents the data and there is no attempt to build a model based upon theory, extrapolation can be disastrous.

As an example of a prediction based upon experimental data, consider one of the earliest sets of experimental data that was analyzed by the method of least squares: the motion of the planets. Johannes Kepler (1571-1630) had access to voluminous amounts of data regarding the motion of the planet Mars that had been recorded by astronomers over many centuries. Kepler used this data to postulate that the motion of the planets were elliptical and not circular as was believed until then. During the early years of the 19th century, both Gauss and Legendre worked independently to develop the method of least squares. The motivation for this work was to use the method to compute the parameters that describe the planetary ellipses. Using the ellipses, we should theoretically be able to predict planetary motion into the future. However, better measurements were made and slight errors were noted. As the time between the calculation of the ellipse parameters and the actual predicted location of a planet increased, the extrapolation error became greater. As the tools of astronomy became more sophisticated, the reason for the extrapolation error became clear: the motion of the planets was not only affected by the gravitational pull of the sun, but also by the gravitational pull of nearby planets. Kepler's elliptical model of planetary motion was a brilliant leap forward in the science of astronomy but the model neglects the very small effects of other planets. As a result, using only the uncorrected elliptic model, the extrapolation into the future led to greater and greater errors.

When there is no theoretical basis for choosing a model, typically one looks at the data and chooses a model that seems reasonable. To illustrate how disastrous this can be when using such a model for extrapolation, the following artificial data set was constructed. Two hundred values of x were generated starting from 0.01 and increasing to 2.00 in increments of 0.01. The values of Y were generated using the following model:

$$Y = 20x^3 - 35x^2 + 10x \tag{3.7.1}$$

The curve generated by this equation is shown in Figure 3.7.1. Looking at only the portion of the curve from 0.01 to 0.5 (i.e., the first 50 points in the data set), one might reasonably assume that the data is represented by a parabola. Using the method of least squares with all 50 points weighted equally, the following parabola is obtained:

$$y = f(x) = -19.700x^2 + 6.848x + 0.1405 \tag{3.7.2}$$

The VR (variance reduction) is a very respectable 99.37% and the RMS error is 0.047. However, when this equation is used to extrapolate to values of x outside the modeling range, we see a very different picture. Results in Table 3.7.1 illustrate this point. In this table the value of σ_f were calculated using Equation 2.6.11. Note in the Y - y column the increasing error in the results. The first point in this table is at x=0.6 and although this point is only slightly outside the modeling range, the error is 0.56 which is about 12 times the **RMS error** noted for the modeling data. For $x = 2$ we see that the computed value of y is -64.964 and this bears no resemblance to the actual value of 40. Note that the estimated standard deviation for this point is only 1.12 while the actual error is almost 105! In other words the actual error is approximately 100 times the estimated standard deviation as computed using Equation 2.6.11.

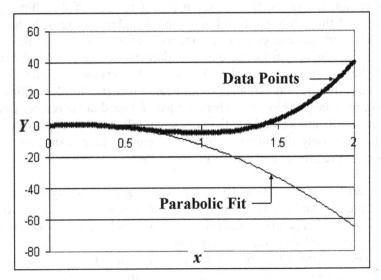

Figure 3.7.1 Cubic Data Fit with a Parabola using Points up to $x = 0.5$

This example illustrates what can happen when a model that seems to adequately estimate Y for values of x within the modeling range is used to extrapolate to values of x outside the range. For this example, as x increases the computed y values become increasingly different then the actual Y values. The conclusion that should be taken from this example is that extrapolation should be avoided whenever one is not certain that the model is firmly based upon theory. Furthermore, the further the value of the independent variable (or variables) is outside the modeling range, the greater is the potential error.

x	Y	y	σ_f	$Y-y$
0.6	-2.28	-2.843	0.040	0.56
0.7	-3.29	-4.719	0.069	1.43
0.8	-4.16	-6.989	0.106	2.83
0.9	-4.77	-9.653	0.149	4.88
1.0	-5.00	-12.712	0.201	7.71
1.1	-4.73	-16.164	0.259	11.43
1.2	-3.84	-20.010	0.325	16.17
1.3	-2.21	-24.250	0.399	22.04
1.4	0.28	-28.885	0.480	29.16
1.5	3.75	-33.913	0.568	37.66
1.6	8.32	-39.335	0.663	47.65
1.7	14.11	-45.151	0.766	59.26
1.8	21.24	-51.361	0.877	72.60
1.9	29.83	-57.966	0.995	87.79
2.0	40.00	-64.964	1.120	104.96

Table 3.7.1 Extrapolation results using Eqs. 3.7.1 and 3.7.2

3.8 Out-of-Sample Testing

For many sets of data there are a large number of data points and thus is it feasible to exclude some of the data from the analysis for subsequent out-of-sample testing. Once a model has been developed, the excluded data is used to determine if the model holds up using the unseen data.

The use of out-of-sample testing is particularly advantageous when the true structure of the model (if there is a true structure) is unknown. For problems in which a number of candidate predictors have been proposed, one might try different combinations in a search for a model that "fits" the data. As an example, consider the problem of attempting to develop a model that predicts the one-day percentage change in the S & P (Standard and Poors) common stock index. The candidate predictors proposed might include various moving averages, volume based parameters, interest rate parameters, etc. One can easily propose tens if not hundreds of candidate predictors for this problem.

The methodology is to break up the data into two (or sometimes three) sets. If the number of candidate predictors is small then two data sets are sufficient. However, if a large number of candidate predictors are being considered, then a third data set is often used. The data sets are typically called the **learning** and **test** data sets and if a third set is used it is called the **evaluation** set. We use the notation *nlrn*, *ntst*, and *nevl* for the number of data records in each of the data sets. Using the *nlrn* learning points, a model is determined using an appropriate modeling technique. The *ntst* test data points are then used with the model and the variance reduction associated with this data set is computed.

The problem with this technique is that if the number of candidate predictors is large, the number of potential models can be huge. As an example, let us again consider the problem of modeling the one-day change in the S & P index and let us assume that 100 candidate predictors are being considered. Let us consider every combination of the candidate predictors up to three dimensions. For every combination we could of course explore many different potential models but to simplify matters let us consider only one model per combination. The total number of models that would have to be considered is all 1 dimensional models (i.e., 100 models), all two dimensional models (i.e., $100*99/2 = 4950$ models) and all three dimensional models (i.e., $100*99*98/6 = 161700$ models) which altogether is 166750 models! Hopefully, some of the models will show significant *VR* (variance reduction) using the *nlrn* learning points and also exhibit significant *VR* using the *ntst* test points. If more than one model is deemed acceptable, one might then consider how to combine models into a single super model.

The problem associated with this procedure is what Aronson calls the **data mining bias** [AR04]. In the S & P index analysis, if over 166 thousand models are considered, isn't there a possibility that a totally irrelevant model might be deemed acceptable purely by chance? To protect against such a possibility, one can use the *nevl* evaluation points as a final test of the power of the model. Failure to pass this final test is fairly conclusive: the objective of the modeling process has not succeeded.

One might ask a simple question: *Why limit the search for a model to a fixed number of dimensions? If there are **ncp** candidate predictors, why not try to build a model with **ncp** dimensions?* This approach is sometimes used by practitioners of neural networks. A weight is associated with each input and the theory states that for irrelevant candidate predictors the computed weights will be close to zero. The problem with this approach is that

the density of the data decreases exponentially with the number of dimensions [WO00]. It is thus a far better strategy to limit the number of dimensions to a value for which the data density is reasonable. Then, if several
sub-spaces are found to have predictive power, use a multi-stage method
for combining the spaces.

The concept of "data density" is not obvious and requires additional consideration. In Figure 3.8.1 we see a desirable distribution of learning data
points for a problem in which there are two independent variables x_1 and
x_2. The points are distributed evenly between the four quadrants of the
data space and cover the range of both x_1 and x_2. A model built using this
data set should perform reasonably well regardless of where a test point
might fall.   In Figure 3.8.2 we see a distribution in which most of the
learning points are located in the upper right quadrant. Note that the number of learning points and the ranges for x_1 and x_2 are the same as in Figure
3.8.1. Yet test points falling in any of the quadrants other than the upper
right quadrant have many fewer nearby neighbors and we would expect
that the predicted values for these test points would be less accurate than
the values for points in the upper right quadrant. Clearly the data density in
the upper right quadrant is much greater than in any of the other quadrants.
Now let us distribute the same number of learning points within a three-
dimensional space. Instead of four quadrants we have 8 distinct regions:
the same four quadrants for x_1 and x_2 but the data in each of these regions
would have to be distributed in two sub-regions: $x_3 \leq x_3avg$ and $x_3 >$
x_3avg. In other words, the average data density in each region is halved
when the number of dimensions in increased by one!

For problems in which the number of candidate predictors is limited, out-
of-sample testing might still be advantageous if many different models are
considered. For example, consider a problem with four candidate predictors and 1000 data points. The purpose of the analysis is to develop a
model to predict performance based upon the candidate predictors. Assume that 30 different model are proposed. Do we use all 1000 points to
develop 30 different models? Then how do we choose which model to
adopt? We could, of course, choose the model exhibiting the greatest variance reduction but is this the best that we can do? For such problems, out-
of-sample testing provides a reasonable alternative. For example, we
might leave 300 to 500 data points out of the initial modeling phase and
then use each model to test how well the model behaves using the test data
set. We might then choose the model exhibiting the best out-of-sample
variance reduction. Once the model has been chosen, then all 1000 points
could be used to recompute the parameters of the chosen model.

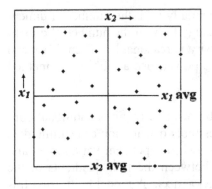 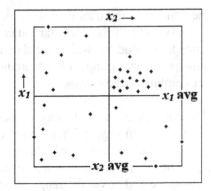

Figure 3.8.1 Desirable Distribution   Figure 3.8.2 Skewed Distribution

In Section 3.7 the subject of extrapolation was considered. For out-of-sample testing one should avoid extrapolation. Any points in the test data set that fall outside the range of data points in the learning data set should be rejected and not considered in the analysis. When considering many different candidate predictors, some points will be included in some subspaces but not in others. There is no need to attempt to find test data sets in which all points fall within the ranges of all the learning point candidate predictors.

It is instructive to consider what we might expect from a model in which there is absolutely no predictive power. The first thought that comes to mind is that the *VR* for the test data set should be zero. It turns out that this is overly optimistic! From equation 3.4.3 the equation for *VR* for the test data set is:

$$VR_{tst} = 100 * \left(1 - \frac{\sum\limits_{i=1}^{i=ntst}(Y_i - y_i)^2}{\sum\limits_{i=1}^{i=ntst}(Y_i - y_{avg})^2} \right) \tag{3.8.1}$$

In this equation, the Y_i's are the actual values and the y_i's are the computed values of y for the test data points. The simplest possible model that we can propose using the learning data set is just the average of the learning values of y. Let us denote this average value as y_{lrn} and the average value of the test data as y_{tst}. In other words, the y_i's in Equation 3.8.1 would simply be y_{lrn} and substituting this into 3.8.1 we get the following:

$$VR_{tst} = 100 * \left(1 - \frac{\sum_{i=1}^{i=ntst}(Y_i - y_{lrn})^2}{\sum_{i=1}^{i=ntst}(Y_i - y_{tst})^2} \right)$$

With a few simple algebraic steps we get the following expression for VR for the test data:

$$VR_{tst} = 100 * \left(1 - \frac{\sum_{i=1}^{i=ntst}(Y_i - y_{tst} + y_{tst} - y_{lrn})^2}{\sum_{i=1}^{i=ntst}(Y_i - y_{tst})^2} \right)$$

$$= -100 * \left(\frac{ntst * (y_{tst} - y_{lrn})^2}{\sum_{i=1}^{i=ntst}(Y_i - y_{tst})^2} \right)$$

(3.8.2)

For models other than just y_{lrn}, we will obtain values of VR distributed about this slightly negative value. We also see in this equation why it is useful to choose the learning and test data sets so that they will be comparable. If there is a significant difference in the average values of Y the VR of the test data might be significantly less than the value obtained for the learning data even if the model is a reasonable representation of data.

To illustrate the points raised in this section, the following artificial data set was generated using the following equation:

$$y = x_1 + 2x_2 - 0.5(x_1 * x_2 + x_1^2 - x_2^2)$$

(3.8.3)

Sixteen combinations of x_1 and x_2 were chosen as shown in Table 3.8.1 with values of y computed using the above equation. The equation used to fit the data was a plane:

$$y = a_1 x_1 + a_2 x_2 + a_3$$

All points were equally weighted. Three separate cases were considered. The data was first analyzed using all 16 points as the learning points. The least squares plane obtained using this data was:

$$y = -2.75x_1 + 3.25x_2 + 3.125 \qquad \text{(Case 1)}$$

For the second case, the first 8 points were used as the learning points and points 9 thru 16 as the test points. The least squares plane obtained for this case was:

$$y = -2.25x_1 + 2.25x_2 + 3.375 \qquad \text{(Case 2)}$$

The third case used points 1, 3, 6, 8, 9, 11, 14 and 16 as the learning points and the remaining eight points as the test points. The least squares plane obtained for this case was:

$$y = -2.75x_1 + 3.25x_2 + 3.000 \qquad \text{(Case 3)}$$

The results for all three cases are summarized in Table 3.8.2. Note that for all three experiments the learning set *VR* is greater than 95%. Note in Case 2 the large deterioration in *VR* of the test data set. This deterioration is due to the large difference in the average values of *y* for the two data sets. In Case 3, the choice of learning set data points resulted in much closer average values of *y* and this resulted in *VR* for the test data set approximately equal to the learning set *VR*.

Point	x_1	x_2	Y
1	1	1	2.5
2	2	1	1.5
3	3	1	-0.5
4	4	1	-3.5
5	1	2	5.5
6	2	2	4.0
7	3	2	1.5
8	4	2	-2.0
9	1	3	9.5
10	2	3	7.5
11	3	3	4.5
12	4	3	0.5
13	1	4	14.5
14	2	4	12.0
15	3	4	8.5
16	4	4	4.0

Table 3.8.1   Values of Y computed using Equation 3.8.3

Case	Learning Points	Test Points	VR_{lrn}	VR_{tst}	y_{lrn}	y_{tst}
1	All 16	None	96.22	N.A.	4.375	N.A.
2	1 -- 8	9 -- 16	95.86	63.55	1.125	7.625
3	1,3,6,8,9,11,14,16	Others	95.41	96.65	4.250	4.500

Table 3.8.2   Results for 3 cases. N.A. is "not applicable"

3.9 Analyzing the Residuals

In Section 3.2 the subject was goodness-of-fit testing based upon estimates of the absolute uncertainties associated with the data. The test involved an analysis of $S / (n-p)$ (the weighted sum of the squares divided by the number of degrees of freedom). In Section 3.3 techniques for selecting the best of several proposed models was discussed. Estimates of absolute uncertainties associated with the data were not necessary in choosing the best of several competing models. In this section our attention is turned to the problem of evaluating a single model in which we do not have knowledge regarding the absolute uncertainties associated with the data.

The runs test is applicable to problems in which the independent variable x is a scalar variable. The basic assumption that is essential for the following analysis is that the residuals are randomly distributed about the resulting least squares curve. After all, if the model is a decent descriptor of the data, we would expect such a random distribution of residuals. This point is illustrated in Figure 3.9.1. The residuals for the 2nd and 8th points are shown for the straight line fit. The residual R_i is defined as $Y_i - y_i$ where y_i is the calculated value of Y_i. Note that R_2 is positive and R_8 is negative.

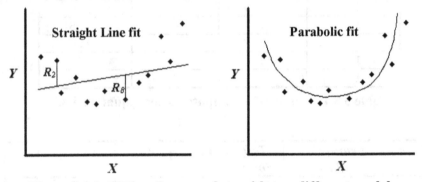

Figure 3.9.1: Fitting the same data with two different models.

We see in this figure that the residuals seem to be randomly distributed about the parabola but the same points are not randomly distributed about the straight line. This is a qualitative observation but what is needed is a quantitative measure of "randomness". The well-known **runs test** is applicable to this task [SI88, DA90, FR92].

The **runs test** is based upon analysis of binary data. When applying this test to the residuals of a least-squares fit, the sign of the residual is a binary indicator (i.e., either positive or negative). In Table 3.9.1 the signs of the residuals are listed for the 13 points for both models. A run is defined as a series of points in which the sign does not change. Examining both Figure 3.9.1 and Table 3.9.1 we see that the first 2 points of the straight line fit are a run of positive points followed by a run of a single negative point, a single positive point, 6 negative points and 3 positive points for a total of 5 runs. Similarly, we note 10 runs for the parabolic fit. Just by looking at Figure 3.9.1 we see that the straight line fit does not adequately represent the data whereas the parabolic fit seems to pass thru the data with randomly distributed residuals. The runs test considers only the number of runs to test the randomness of the residuals.

Point i	R_i (line)	Runs (line)	R_i (parabola)	Runs (parabola)
1	+	1	–	1
2	+	1	+	2
3	–	2	–	3
4	+	3	+	4
5	–	4	–	5
6	–	4	–	5
7	–	4	+	6
8	–	4	–	7
9	–	4	+	8
10	–	4	+	8
11	+	5	+	8
12	+	5	–	9
13	+	5	+	10

Table 3.9.1   Residuals and *runs* for data seen in Figure 3.9.1

The test is based upon calculation of the probability of observing a number of runs less than or equal to a given limit. For example, what is the probability of observing 5 or less runs for the straight line model and 10 or less runs for the parabolic model? If *runs* is an even number than the number of both positive and negative runs is equal to *k* which must be exactly *runs* / 2. If *runs* is an odd number than one of the groups will have *k* runs and the other *k*-1 runs where *runs* = 2*k*-1. The assumption of randomness implies that all possible orderings of the residuals are equally probable. If there are *N* data points and n_1 of the first type and n_2 of the second type, the total number of combinations of orderings is $C_{n_1}^N$ (where $N = n_1 + n_2$).

(Note that $C_{n_1}^N$ is exactly equal to $C_{n_2}^N$). For example, if n_1=3 (number of +'s) and n_2 = 2 (number of –'s), the number of combinations is C_3^5 = 5*4*3/(3*2*1) = 10. This is exactly equal to C_2^5 = 5*4/(2*1). The ten possible orderings are shown in Table 3.9.2. Note that there are two combinations with 2 runs, three with 3 runs, four with 4 runs and one with 5 runs. Let us first consider the case of an even number of runs (e.g., *runs* = 4 and therefore *k* = 2). There are two possible partitions of the 3 +'s: ++ | + and + | ++. There is only one way to partition the 2 –'s: - | -. In general we want to place *k*-1 bars (i.e., | symbols) in the n_1 - 1 gaps between the +'s.

The number of partitions is therefore $C_{k_1-1}^{n_1-1}$ For example, if n_1=4 and
k=3, C_2^3 =3 and the three partitions are: +|++|+. ++|+|+ and +|+|++. We are
now in a position to state the probability of achieving *runs* for N data
points when N is an even or odd number:

$$Prob(runs) = \frac{2C_{k-1}^{n_1-1}C_{k-1}^{n_2-1}}{C_{n_1}^N} \quad runs \text{ even, } k = runs /2 \quad (3.9.1)$$

$$Prob(runs) = \frac{C_{k-1}^{n_1-1}C_{k-2}^{n_2-1} + C_{k-1}^{n_2-1}C_{k-2}^{n_1-1}}{C_{n_1}^N} \quad (3.9.2)$$

$$runs \text{ odd, } 2k-1=runs$$

The 2 in the numerator of Equation 3.9.1 is required because the runs can
start with either a positive or negative run. From Table 3.9.2 we see that
the probability of observing exactly 4 runs is 0.4 (i.e., 4 out of 10 of the
combinations have 4 runs). Checking this with Equation 3.9.1:

$$Prob(4) = \frac{2C_1^2C_1^1}{C_3^5} = \frac{2*2*1}{10} = 0.4 \quad (3.9.3)$$

Combination	Order	Runs
1	+++--	2
2	--+++	2
3	+--++	3
4	++--+	3
5	-+++-	3
6	+-++-	4
7	++-+-	4
8	-++-+	4
9	-+-++	4
10	+-+-+	5

Table 3.9.2   All possible orderings for n_1=3 and n_2=2.
n_1 is number of +'s, n_2 is number of -'s

Equations 3.9.1 and 3.9.2 are used and tabulations for combinations of n_1 and n_2 are included in many books. For example, Freund [FR92] includes a table that contains 2.5% lower and upper bounds for all combinations up to n_1 and $n_2 = 15$. Siegel and Castellan [SI88] include a table that extends the range up to n_1 and $n_2 = 20$. If n_1 and n_2 are greater than or equal to 10, the normal approximation can be used. The mean and variance of the distribution is:

$$\mu = \frac{2n_1 n_2}{n_1 + n_2} + 1 \qquad (3.9.4)$$

$$\sigma^2 = \frac{2n_1 n_2 \left(2n_1 n_2 - n_1 - n_2\right)}{\left(n_1 + n_2\right)^2 \left(n_1 + n_2 - 1\right)} \qquad (3.9.5)$$

When analyzing residuals, if they are randomly distributed it is a reasonable assumption that the values of n_1 and n_2 are approximately equal to $N/2$ so the equations reduce to the following:

$$\mu = \frac{N}{2} + 1 \qquad (3.9.6)$$

$$\sigma^2 = \frac{N\left(N/2 - 1\right)}{2\left(N - 1\right)} \qquad (3.9.7)$$

For example, assume that $N = 40$ and we observe 13 runs. What is the probability of observing this number of runs if the residuals are randomly distributed? The mean value μ of the distribution is 21 and σ is $(20*19/39)^{1/2} = 3.12$. The probability of observing 13 runs is the area under the normal curve from $x = 12.5$ to $x = 13.5$. We can convert these values to the standard normal curve by subtracting μ and dividing by σ. The range is therefore:

$$\frac{12.5 - 21}{3.12} = -2.724 \leq u \leq \frac{13.5 - 21}{3.12} = -2.404$$

From a table of the standard normal distribution 2.724 corresponds to a probability of 0.49675 and 2.404 corresponds to a probability of 0.4919 so the probability of falling within this range is 0.49675 − 0.4919 = 0.00485.

We can also compute the probability of observing 13 or less runs. The probability is $0.5 - 0.4919 = 0.0081$ which is less than 1%. For values of $N < 20$, the normal approximation is not recommended. Values of the maximum number of runs for rejecting the randomness hypotheses computed using Equations 3.9.1 and 3.9.2 are included in Table 3.9.3. For even values of N, n_1 and $n_2 = N/2$. For odd values of N, $n_1 = n_2 \pm 1$. The *runs* limits in the table are based upon a confidence limit of 2.5%. In other words, if the residuals are randomly distributed, the probability of observing the values listed in the table or less is less than or equal 2.5% (i.e., Prob(*runs* $\leq$ *runs limit*) ≤ 0.025).

N	runs limit		N	runs limit
9	2		17	5
10	2		18	5
11	3		19	5
12	3		20	6
13	3		21	6
14	3		22	7
15	4		23	7
16	4		24	7

Table 3.9.3   2.5% *limit* for *runs*. (Reject randomness hypothesis if *runs* $\leq$ *runs_limit*)

As an example of the use of the **runs test**, consider the data in Figure 3.9.2. The data was tested using the REGRESS program with four different models. No information was available regarding the uncertainties associated with the values of Y so unit weighting was used. For this reason the resulting values of $S / (n\text{-}p)$ are only meaningful on a relative basis. The resulting values for $S / (n\text{-}p)$ and *runs* are included in Table 3.9.4. We see from the results in the table that Model 1 is best on the basis of both $S / (n\text{-}p)$ and *runs* but looking only at $S / (n\text{-}p)$, we cannot conclude whether or not this model is an adequate representation of the data. However, the value of *runs* confirms that the residuals are randomly distributed about the least squares curve and therefore the model passes this goodness-of-fit test. Notice the value of *runs* for Model 1 is very close to the expected average of 51 as computed using Equation 3.9.6. The *runs* values for the other models are well below the 2.5% confidence limit of 40. The REGRESS output for Models 1 and 2 is shown in Figure 3.9.3.

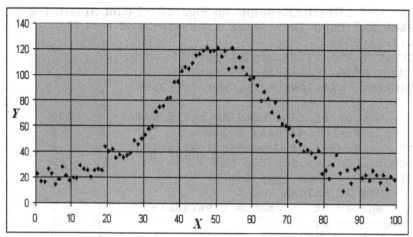

Figure 3.9.2: Data analyzed with 4 different models (Table 3.9.4)

Model	Equation	$S/(n-p)$	Runs
1	$A1*EXP(-((X - A3) / A2)^2) + A4$	23.27	50
2	$A1*EXP(-((X - A3) / A2)^2)$	86.84	24
3	$A1 + A2*X + A3*X^2$	362.83	10
4	$A1 + A2*X + A3*X^2 + A4*X^3$	366.41	10

Table 3.9.4   Results for 4 different models for data in Fig 3.9.2

Figure 3.9.3   REGRESS output for Figure 3.9.2 data, Models 1 & 2

```
PARAMETERS USED IN REGRESS ANALYSIS: Wed Nov 17
14:42:31 2004
  INPUT PARMS FILE: fig392.par
  INPUT DATA  FILE: fig392.dat
  REGRESS  VERSION: 4.10, Nov 15, 2004

  N - Number of recs used to build model  :    100
  NCOL - Number of data columns           :      4
  NY   - Number of dependent variables    :      1
  YCOL1 - Column for dep var Y                 :  2
  SYTYPE1 - Sigma type for Y              :      1
     TYPE 1:  SIGMA Y = 1
  M - Number of independent variables     :      1
  Column for X1                           :      1
  SXTYPE1 - Sigma type for X1             :      0
     TYPE 0:  SIGMA X1 = 0
```

Analysis for Model 1
  Function Y: A1*EXP(-((X-A3)/A2)^2)+A4

K	A0(K)	AMIN(K)	AMAX(K)	A(K)	SIGA(K)
1	50.00000	Not Spec	Not Spec	99.15170	1.36402
2	10.00000	Not Spec	Not Spec	19.97158	0.38467
3	40.00000	Not Spec	Not Spec	50.19843	0.19424
4	0.00000	Not Spec	Not Spec	20.22799	0.95556

```
Variance Reduction:           98.22
S/(N - P)           :         23.27434
RMS (Y - Ycalc)     :          4.72688
Runs Test: Number runs = 50 Must be > 40 to pass test.
This limit is based upon 2.5% confidence level.
Expected avg number of runs if residuals random: 51.0.
```

```
Analysis for Model 2
  Function Y:  A1*EXP(-((X-A3)/A2)^2)

    K      A0(K)     AMIN(K)     AMAX(K)       A(K)      SIGA(K)

    1   50.00000   Not Spec   Not Spec   111.39400     1.96153
    2   10.00000   Not Spec   Not Spec    27.18914     0.55796
    3   40.00000   Not Spec   Not Spec    50.17885     0.39033

Variance Reduction:            93.27
S/(N - P)             :         86.83555
RMS (Y - Ycalc)       :          9.17772
Runs Test: Number runs = 24 Must be > 40 to pass test.
This limit is based upon 2.5% confidence level.
Expected avg number of runs if residuals random: 51.0.
```

Figure 3.9.3 (continued)   REGRESS output: Model 2

Chapter 4 CANDIDATE PREDICTORS

4.1 Introduction

For many experimental data sets, we do not have a well-defined mathematical model nor are we sure which independent variables should be included in the model. Examples of such problems can be found in econometrics, medicine, agriculture and many other areas of science and engineering. We use the term **candidate predictors** to classify potential variables that might or might not appear in the final model. The selection of the candidate predictors is of course problem dependent. The group collecting the data has knowledge of the problem area and can usually suggest variables that should be considered as candidates for inclusion in a model.

Consider as an example, the following problem in the field of pharmacology. A drug company wishes to develop a model for predicting the effectiveness of a drug to reduce blood pressure. Clearly, the amount of the drug given to each patient is an important candidate for predicting the drug effectiveness and will most certainly appear in the model. Other candidate predictors should include variables related to the patient: blood pressure, age, weight, variables related to his or her medical history, etc. The drug itself might include some variability in its chemical composition and this variability might be included as additional candidate predictors. One can see that the number of candidate predictors can rise rapidly. The greater the number of candidate predictors, the greater the probability that an important predictor is not overlooked. However, as the number of candidate predictors rises, the difficulty in finding the best model also rises.

A second example is taken from the field of econometrics. Consider the problem of attempting to develop a model for predicting the change in the United States unemployment percentage over the next month. Vast amounts of data are collected to measure many variables related to the performance of the U.S. economy and some of this data might be relevant for

the desired model. For example, changes in the unemployment percent-
ages over the past months would certainly be prime candidate predictors.
Other changes might also be relevant: changes in the number of housing
starts, changes in retail sales, changes in the gross national product, etc.
Other measures of the strength of the economy might include changes in
the stock market indices, changes in interest rates, etc. Measures related to
the world-wide economy might also be included in the list of candidate
predictors. Once again, the list of candidate predictors can grow as the
analyst considers a widening range of variables that might or might not affect the rate of unemployment.

Before attempting to build a model, the analyst should attempt to answer
some basic questions regarding candidate predictors. Specifically can we
say anything regarding the worth of a particular candidate predictor? Are
some candidate predictors redundant? Do some subsets of the **candidate
predictor space** contain more information than others? In this chapter
some useful tools for answering such questions are considered. If the set
of candidate predictors can be reduced to a subset with greater potential for
being included in the final model, the entire modeling process will be sim-
plified and the potential for success will be increased. In the statistical literature the term **measures of association** is used to describe tools that
have been developed for measuring the dependence of one variable upon
another [e.g., SI88, DA90]. In this chapter some of the most powerful and
useful measures of association are discussed and examples are included to
illustrate how they can be applied to specific problems.

4.2 Using the F Distribution

In Section 1.3 the F distribution was discussed. We can use the F distribution to answer the following question: Does a particular subspace contain
information related to a dependent variable y? The term "subspace"
applies to subspaces of the candidate predictor space. Each candidate predictor is a single dimension in the larger ncp (number of candidate predictors) space. If a particular subspace does not have "predictive power" it
does not necessarily imply that it should be immediately rejected from further consideration. It might, when combined with another subspace, be a
powerful predictor. However, when looking for ways to reduce ncp, one
tends to choose candidate predictors which on their own as one-
dimensional spaces or combined with other candidate predictors show that
they do contain information regarding y.

To analyze a subspace using the **F** distribution, the space is partitioned into **cells.** The number of cells is set by the analyst and should be based upon the number of available data points. Clearly, the greater the number of data points, the greater can be the number of cells. The number of dimensions in the subspace is irrelevant. We see cells for one and two dimensional spaces in Figures 4.2.1 and 4.2.2. Note that there is no need to make the cell sizes constant. A worthwhile strategy is to choose the cell sizes such that the cells are approximately equally populated. To accomplish this for one dimensional spaces, the data is first sorted and then the cell dimensions are chosen to achieve approximate equality. For example if 1000 data points are available and we wish to partition the data into 10 cells, we would sort the data so that x_1 is the smallest value of x and x_{1000} is the largest value. The upper limit of cell 1 would be $(x_{100} + x_{101}) / 2$, the upper limit of cell 2 would be $(x_{200} + x_{201}) / 2$, etc. One might ask: *What happens if there are duplicates?* For example, if x_{99} thru x_{105} have exactly the same value? The cell limit would be this value and points 99 and 100 would be included in cell 1 and points 101 thru 105 would be in cell 2.

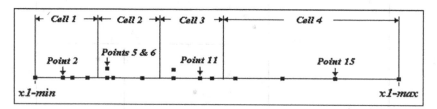

Figure 4.2.1   Partitioning a one dimensional space with 16 points. Notice that each cell is equally populated. Notice points 5 & 6 and 9 & 10 have same values.

For two or more dimensional spaces, the selection of the cell sizes can be more complicated. First of all, if we want to compare spaces of varying dimensions, it is statistically reasonable to try to maintain equally populated cells over all spaces. From the preceding discussion of the one-dimensional partition, how would we proceed to 10 equally populated two dimensional cells? A general solution to this problem for **p** dimensions is included in my last book [WO00]. A data structure called the **p-tree** is described that is used to generate cell limits when the number of cells is 2^h (where **h** is the tree height). Thus this data structure cannot be used to divide a space into 10 cells but it could be used to divide it into 8 (i.e. **h**=3) or 16 (i.e. **h**=4) cells. Figure 4.2.2 is an example of a two dimensional space divided into 8 cells. In this example the number of data points per cell is 2. For real problems we would want a much larger cell density. The **p-tree** is a generalization of a two-dimensional data structure called a **quadtree**. The quadtree, is used extensively in computer graphics, computer aided design, image processing, etc. [SA90].

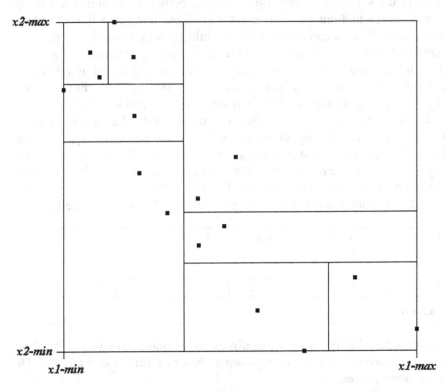

Figure 4.2.2   Partitioning data in x_1-x_2 space. Notice that each cell is equally populated with 2 points per cell.

One might ask the question: For a two dimensional space, why not just sort both dimensions into r equally populated regions creating r^2 cells? Similarly, for a three dimensional space, why not just sort each dimensions into r equally populated regions creating r^3 cells? To maintain equality in average cell density over varying dimensional spaces, we would have to choose different values of r for different numbers of dimensions. For example, if we used $r = 64$ for one dimensional spaces, $r = 8$ for two dimensional spaces, and $r = 4$ for three dimensional spaces, the number of cells for these spaces would be 64 and the average cell density would be the same regardless of the dimensionality up to three dimensions. This approach is reasonable if the values of the x's in the different dimensions are uncorrelated. However, if there is a measurable level of correlation, the cells would certainly not be equally populated. For a two dimensional space, the problem using this technique is illustrated in Figure 4.2.3. This figure is based upon the same data as used in Figure 4.2.2 where the values of x_1 and x_2 are clearly correlated. Using $r = 4$, there are a total of 16 data points and 16 cells so the average cell density is 1. For this example, this partitioning scheme leaves 9 cells empty and one cell with four points. As the correlation increases, the disparity between the most and least populated cells increases. It should be emphasized that the correlation need not be linear. The concept of nonlinear correlation is introduced in the next section. This problem is avoided by using the p-tree data structure for partitioning the space.

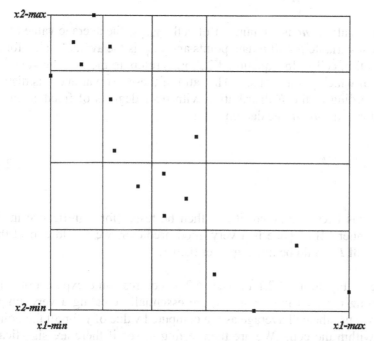

Figure 4.2.3   Partitioning data in x_1-x_2 space. Each dimension is partitioned independently. Notice that in each row and in each column there are 4 points.

Once we have divided our data into cells, the first step in the analysis is to compute the mean and variance of the dependent variable over all cells. We then compute the mean and variance within each cell. We can compute the variance between cells σ_b^2 and the variance within cells σ_w^2 as follows:

$$\sigma_b^2 = \frac{\sum_{j=1}^{j=nc} n_j (y_{avg,j} - y_{avg})^2}{nc - 1} \tag{4.2.1}$$

$$\sigma_w^2 = \frac{\sum_{i=1}^{i=n} (y_i - y_{avg,j})^2}{n - nc} \tag{4.2.2}$$

In these equations nc is the number of cells, y_{avg} is the average value of the dependent variable for all n data points and $y_{avg,j}$ is the average value for all points within cell j. In Equation 4.2.2, $y_{avg,j}$ refers to the average value for the cell in which point i falls.   The ratio of these two variances is distributed according to the F distribution with nc-1 degrees of freedom in the numerator and n-nc in the denominator:

$$F = \frac{\sigma_b^2}{\sigma_w^2} \tag{4.2.3}$$

Once F has been computed, it can then be tested for significance in the usual manner. If a space is a very good predictor, we would expect that the value of F would be much greater than 1.

The use of Equations 4.2.1 through 4.2.3 requires some explanation. By partitioning the data into cells, we are essentially creating a very simple model: we use the cell average as the computed value of y for all the points falling within the cell. We are then testing to see if there are significant differences in the average values of y in the different regions (i.e., cells) of the space. To illustrate this process consider the data in Table 4.2.1. We have partitioned 200 data points into four cells, each cell containing 50 points. (It should be emphasized that it is not essential that the number of points per cell be the same for all cells.) The table includes the average value of y and s (the unbiased estimate of σ) for each cell.   Note that no mention is made regarding the dimensionality of this space. It could be 4 cells in a one dimensional or a two dimensional space. (To partition data in a three dimensional space we would need at least 8 cells.)

j	$y_{avg,j}$	n_j	s_j
1	12.0	50	2.0
2	4.0	50	1.0
3	5.0	50	1.5
4	11.0	50	3.0

Table 4.2.1   Data Partitioned into 4 Cells

The average value y_{avg} for this data is $(12+4+5+11) / 4 = 8$. To compute σ_b^2 we use Equation 4.2.1:

$$\sigma_b^2 = 50 * ((12-8)^2 + (4-8)^2 + (5-8)^2 + (11-8)^2)/(4-1)$$
$$= 2500/3 = 833.33$$

We can use the following equality to compute σ_w^2:

$$(n-nc)\sigma_w^2 = \sum_{j=1}^{j=nc}(n_j-1)s_j^2 \qquad (4.2.4)$$

$$\sigma_w^2 = 49 * (2^2 + 1^2 + 1.5^2 + 3^2)/(200-4)$$
$$= 796.25/196 = 4.0625$$

The value of F is $833/4.06 = 205$ which is a very large number. The 1% confidence limit for F (that is, $F(0.01, 3, 196)$) is about 3.9 so 205 is orders of magnitude above this limit. This result should not be surprising as we see that the average values of y in cells 1 and 4 are much larger than the values in cells 2 and 3 indicating that this space is a true predictor for y.

4.3 Nonlinear Correlation

Shannon introduced the concept of nonlinear correlation in his ground-breaking paper *A Mathematical Theory of Communication* [SH48]. This paper was published by the Bell Laboratories in 1948 and has had far-reaching consequences. Nonlinear correlation was only one of several important concepts and ideas included in the paper. In this paper Shannon described a method for measuring nonlinear correlation on a scale from zero to one. A value of zero implies that two variables are unrelated: for example, knowledge of v_1 provides no knowledge of v_2. A value of one implies that the two variables are completely related: for example, knowledge of v_1 provides exact knowledge of the value of v_2 and knowledge of v_2 provides exact knowledge of the value of v_1.

Nonlinear correlation can be used to accomplish several tasks. It can be used to determine whether an independent variable x (perhaps one of many candidate predictors) contains useful information for building a model for a dependent variable y. It can also be used to determine if candidate predictors x_1 and x_2 are related. If, for example, there is a high degree of nonlinear correlation between these two candidate predictors, then one of them might be eliminated from the set of candidate predictor variables.

To understand the concept of nonlinear correlation consider an experiment in which 100 data points are observed and for each data point values of both x_1 and x_2 are recorded. The values of both x_1 and x_2 are integers from 1 to 5, so there are 25 possible combinations. Consider the distributions of points shown in Figures 4.3.1 through 4.3.3.

In Figure 4.3.1 we see that if the data is truly indicative of the population, then all we need to know is the value of x_1 and we would automatically know the value of x_2. For example, if x_1 is 1, then x_2 is 3 and if x_1 is 2, then x_2 is 5, etc. Conversely, knowledge of x_2 yields the value of x_1. For example, if x_2 is 2, then x_1 is 4. The nonlinear correlation coefficient (CC) for this example is one. Clearly, if we increase the number of data points beyond 100, we might discover that cells unpopulated by the initial 100 data points become populated. If this happens, then a recalculation of CC would yield a value less than one. But based only upon the information contained in the first 100 points, our best estimate of CC is one.

In Figure 4.3.2 we see an entirely different picture. Knowledge of x_1 does not add to our knowledge of x_2. The most probable value of x_2 is 2 (i.e., 30% of all data points have a value $x_2 = 2$). We see, however, that this percentage is true for all values of x_1. (For $x_1 = 1$ and 4, 9 out of 30 data points, for $x_1 = 2$, 6 out of 20 data points and for $x_1 = 3$ and 5, 3 out of 10 data points have values of $x_2 = 2$.) Similarly, knowledge of x_2 does not add to our knowledge of x_1. For this case, CC is thus zero.

	$x2=1$	$x2=2$	$x2=3$	$x2=4$	$x2=5$
$x1=1$			5		
$x1=2$					35
$x1=3$	15				
$x1=4$		25			
$x1=5$				20	

Figure 4.3.1   Distribution of 100 data points. This distribution exhibits nonlinear correlation $CC = 1$.

	x2=1	x2=2	x2=3	x2=4	x2=5
x1=1	6	9	6	6	3
x1=2	4	6	4	4	2
x1=3	2	3	2	2	1
x1=4	6	9	6	6	3
x1=5	2	3	2	2	1

Figure 4.3.2   Distribution of 100 data points. This distribution exhibits nonlinear correlation $CC = 0$.

	x2=1	x2=2	x2=3	x2=4	x2=5
x1=1		10			10
x1=2		10			
x1=3	10		5		5
x1=4				20	
x1=5	20			10	

Figure 4.3.3   Distribution of 100 data points. This distribution exhibits nonlinear correlation CC between zero and one.

In Figure 4.3.3 we see that knowledge of x_1 does add to our knowledge of x_2 and visa versa. For example, if $x_1 = 1$ then there is a 50% probability that $x_2 = 2$ and a 50% probability that $x_2 = 5$. If $x_1 = 2$ then there is a 100% probability that $x_2 = 2$. If $x_1 = 3$ then there is a 50% probability that $x_2 = 1$ and a 25% probability that $x_2 = 3$ or 5. If $x_1 = 4$ then there is a 100% probability that $x_2 = 4$. And finally, if $x_1 = 5$ two-thirds of the values of x_2 will be 1 and one-third will be 4. For this case *CC* is between zero and one.

For continuous data one must partition the data into cells. The partitioning scheme shown in Figure 4.2.3 rather than in Figure 4.2.2 is preferable. For the purpose of examining correlation, equally populated cells are of no interest. However, if each variable is divided into equally populated bins, then our partitioning is optimum (in the sense of maximizing entropy or uncertainty) [PY99]. The term *entropy* was taken from thermodynamics and was applied by Shannon to the field of information theory. We can consider the information content of a particular value (or cell) of a variable as $-p·log_2(p)$ where p is the probability of the variable taking on the value. (The minus sign is required because $log_2(p)$ is negative.) An explanation of this formulation for information content is included in Pyle's book on Data Mining [PY99]. As an example of the information content of a cell, consider the variable x_1 in Figure 4.3.3. The probability that its value is 1 is 0.2 (i.e., 20 out of 100 data points have the value $x_1 = 1$). The information content for $x_1 = 1$ is thus $-0.2·log_2(0.2)$. Using the notation of $H(v)$ to denote the entropy associated with the variable v, we compute the entropy as the sum of the information content for all possible values of v:

$$H(v) = -\sum_{i=1}^{i=r} p_i \log_2 (p_i)$$

(4.3.1)

In this equation r is the number of values that v can assume and p_i is the probability that it assumes the ith value. Note that the sum of all p_i's must equal 1:

$$\sum_{i=1}^{i=r} p_i = 1$$

(4.3.2)

To understand why equally populated cells maximize entropy, consider the 5 cases shown in Table 4.3.1. In this table the probabilities of v assuming 4 different values are listed for each case. Note that case 1 exhibits maximum entropy (maximum uncertainty) and case 5 exhibits minimum entropy. In case 5 there is no uncertainty: the value of v is always one. As examples of the calculations, the value of $log_2(0.25) = -2$ so $H(v)$ for case 1 is $4*0.25*2 = 2.00$. For case 3 the value of $log_2(0.5) = -1$ so $H(v)$ is $0.5 + 0.5 + 0 + 0 = 1.00$. Comparing cases 1 and 2 we see that case 2 is slightly more informative and therefore its entropy is slightly less.

case	v=1	v=2	v=3	v=4	H(v)
1	0.25	0.25	0.25	0.25	2.000
2	0.30	0.20	0.25	0.25	1.985
3	0.50	0.50	0.00	0.00	1.000
4	0.60	0.40	0.00	0.00	0.971
5	1.00	0.00	0.00	0.00	0.000

Table 4.3.1   Entropy calculations for 5 different cases

When we are interested in computing the nonlinear CC associated with two variables, we compute the entropy of each variable separately and then combined. For example, if we wish to compute CC for variables x_1 and x_2 we must first compute 3 entropies. The entropies for x_1 and x_2 computed separately are:

$$H(x_1) = -\sum_{i=1}^{i=r1} p_i log_2(p_i) \quad \& \quad H(x_2) = -\sum_{j=1}^{j=r2} p_j log_2(p_j) \qquad (4.3.3)$$

In these equations r_1 and r_2 can be different. A difference might occur if x_1 and x_2 are discrete variables with different ranges. However, if they are continuous variables, they must be "discretized" by dividing each variable into separate regions or bins and typically the value of r_1 and r_2 would be chosen to be the same. We compute the entropy associated with the two dimensional partitioning of the data as follows:

$$H(x_1 x_2) = -\sum_{i=1}^{i=r1} \sum_{j=1}^{j=r2} p_{ij} log_2(p_{ij}) \qquad (4.3.4)$$

We are now in a position to compute CC:

$$CC = \frac{2(H(x_1) + H(x_2) - H(x_1 x_2))}{H(x_1) + H(x_2)} \quad (0 \le CC \le 1) \qquad (4.3.5)$$

As an example, consider the data shown in Figure 4.3.1. We can present this data in the form of probabilities as shown in Table 4.3.2:

	$x_2=1$	$x_2=2$	$x_2=3$	$x_2=4$	$x_2=5$	Row Sum
$x_1=1$			0.05			0.05
$x_1=2$					0.35	0.35
$x_1=3$	0.15					0.15
$x_1=4$		0.25				0.25
$x_1=5$				0.20		0.20
ColumnSum	0.15	0.25	0.05	0.20	0.35	1.00

Table 4.3.2   Recasting Figure 4.3.1 into probabilities and summing rows and columns.   Note that the sum of all rows and all columns is one.

The values of all three entropies (i.e., $H(x_1)$, $H(x_2)$, and $H(x_1 x_2)$) are equal to 2.121:

$$H = -(0.15 \, log_2 (0.15) + 0.25 \, log_2 (0.25) + ... + 0.35 \, log_2 (0.35)) = 2.121$$

Substituting into Equation 4.3.5 we see that CC is indeed 1:

$$CC = \frac{2(2.121 + 2.121 - 2.121)}{2.121 + 2.121} = 1$$

In a similar manner we can show that CC for the data in Figure 4.3.2 is zero. Now let us compute CC for the data in Figure 4.3.3. The data in Figure 4.3.3 is presented as probabilities as shown in Table 4.3.3:

	$x_2=1$	$x_2=2$	$x_2=3$	$x_2=4$	$x_2=5$	Row Sum
$x_1=1$		0.10			0.10	0.20
$x_1=2$		0.10				0.10
$x_1=3$	0.10		0.05		0.05	0.20
$x_1=4$				0.20		0.20
$x_1=5$	0.20			0.10		0.30
ColumnSum	0.30	0.20	0.05	0.30	0.15	1.00

Table 4.3.3   Recasting Figure 4.3.3 into probabilities and summing rows and columns. Note that the sum of all rows and all columns is 1.

The three entropies are computed as follow:

$$H(x_1) = -(3 * 0.2 \, log_2(0.2) + 0.1 \, log_2(0.1) + 0.3 \, log_2(0.3)) = 2.246$$

$$H(x_2) = -(2 * 0.3 \, log_2(0.3) + 0.2 \, log_2(0.2) + 0.0.05 \, log_2(0.0.05) +$$
$$0.0.15 \, log_2(0.0.15)) = 2.133$$

$$H(x_1 x_2) = -(5 * 0.1 \, log_2(0.1) + 2 * 0.2 \, log_2(0.2) +$$
$$2 * 0.0.05 \, log_2(0.0.05)) = 3.022$$

The resulting value of **CC** (from Equation 4.3.5) is 0.620. This value implies a high degree of nonlinear correlation and implies that knowledge of x_1 helps us predict the value of x_2 and visa versa.

Can we say anything about the significance of **CC**? If there is no connection between x_1 and x_2 (the null hypothesis) what is the probable range of values that we should expect? We define a new variable T as follows:

$$\begin{aligned} T &= 2N(H(x_1) + H(x_2) - H(x_1 x_2)) \\ &= N * CC * (H(x_1) + H(x_2)) \end{aligned} \tag{4.3.6}$$

The variable T approaches a χ^2 distribution with ν degrees of freedom as N becomes large [RA73]:

$$\nu = (r_1 - 1)(r_2 - 1) \tag{4.3.7}$$

x_1	x_1-bin	x_2	x_2-bin	Y	y-bin
0.3190	2	0.0368	1	0.7214	4
0.7537	4	0.0397	1	0.0586	2
0.7792	4	0.4907	2	0.5695	2
0.1009	1	0.4552	1	0.3921	1
0.3132	2	0.7657	3	0.7186	4
0.8893	4	0.8450	3	0.4974	1
0.6939	3	0.8685	4	0.6245	3
0.0720	1	0.4763	2	0.3007	1
0.4743	2	0.8673	4	0.7313	4
0.8069	4	0.8278	3	0.5513	2
0.6763	3	0.8457	4	0.6355	3
0.7189	3	0.8264	3	0.6087	3
0.1071	1	0.9068	4	0.4099	1
0.2513	1	0.4275	1	0.6721	3
0.3770	2	0.5003	2	0.7389	4
0.8297	4	0.7338	3	0.5363	2
0.5519	3	0.6293	2	0.7033	3
0.5177	2	0.8470	4	0.7175	4
0.7256	3	0.2964	1	0.6044	2
0.0021	1	0.5832	2	0.0107	1

Table 4.3.4   20 Data points and associated bins

As an example of the use of CC as a precursor to modeling, consider the data in Table 4.3.4. The table includes actual values of x_1, x_2 and y and "binned" values from 1 to 4. The bin values were assigned by first sorting the variable and then assigning a value of one to the first quarter, two to the second quarter, etc. For example, all values of $x_1 \leq 0.2513$ were assigned bin values of one, all values of $x_1 > 0.2513$ and $x_1 \leq 0.5177$ were assigned values of two, etc. The number of data points in each bin is equal to n (the number of data points) divided by nb (the number of bins) which is $20/4 = 5$. The number of cells is $nb * nb = 16$. The average cell density is very small ($20/16 = 1.25$) but the purpose of this example is to just illustrate the procedure. What we are interested in determining is if either x_1 or x_2 or both are useful independent variables for a model for y. Note that the scheme used to divide the data into bins creates 4 equally populated bins for each variable. So from Case 1 of Table 4.3.1 the entropies $H(x_1)$, $H(x_2)$ and $H(y)$ are all exactly 2.00.

The first question to be answered is whether or not x_1 and x_2 are correlated (in the nonlinear sense). To compute CC we therefore need the value of

$H(x_1x_2)$. We can create a table similar to Table 4.3.3 from the data in 4.3.4:

	$x_2=1$	$x_2=2$	$x_2=3$	$x_2=4$	Row Sum
$x_1=1$	0.10	0.10		0.05	0.25
$x_1=2$	0.05	0.05	0.05	0.10	0.25
$x_1=3$	0.05	0.05	0.05	0.10	0.25
$x_1=4$	0.05	0.05	0.15		0.25
ColumnSum	0.25	0.25	0.25	0.25	1.00

Table 4.3.5   Recasting the data from Table 4.3.4 into probabilities

From this table we see that the value of $H(x_1x_2)$ is 3.684:

$$H(x_1 x_2) = -(4 * 0.1 \, log_2(0.1) + 0.15 \, log_2(0.15) +$$
$$9 * 0.0.05 \, log_2(0.0.05)) = 3.684$$

Using Equation 4.3.5 the computed value of CC is 0.158 and we next con-sider whether or not this value is significant. From Equations 4.3.6 and 4.3.7 the value of $T = 12.6$ and the number of degrees of freedom $v = 9$. The 5% confidence limit for the χ^2 distribution with 9 degrees is 16.9 so this value in not significant at this confidence level. We can conclude that there is no reason to assume that x_1 and x_2 are correlated (in the nonlinear sense).

We next turn our attention towards the relationship between x_1 and y. In a similar manner the entropy $H(x_1y)$ is computed to be 2.541 and CC is 0.729. This value of CC appears to be quite high but does it pass the test of significance? The value of T is 58.3 and this number is far beyond the 5% confidence limit. It is also far beyond the 1% confidence limit (21.7) so we can conclude the x_1 and y are definitely related. Is y also related to x_2? The entropy $H(x_2y)$ is computed to be 3.821 and CC is 0.089. This value is even less than 0.159 computed for x_1 and x_2 so we conclude that it is also not significant and there is no evidence that a relationship between x_2 and y exists. Perhaps with many more data points we might be able to detect a significant (but weak) relationship between these 2 variables but with the available evidence we can only conclude that at this point we can-not detect a significant effect. We should, however, note that this analysis does not exclude the possibility that a two dimensional model (i.e., y as a function of both x_1 and x_2) might be a better model then y as just a function of x_1.

What if we had detected the same value of CC = 0.089 between x_2 and y but based upon 1000 data points after dividing each variable into 10 equally populated bins? Would this value be significant? The values of $H(x_2)$ and $H(y)$ would be 3.322:

$$H(x_2) = H(y) = -(10 * 0.1 \, log_2 (0.1)) = - log_2 (0.1) = 3.322$$

The value of T would be 591:

$$T = n * CC * (H(x_2) + H(y)) = 1000*0.089 * (3.322 + 3.322) = 591$$

The number of d.o.f. (degrees of freedom) is 81. The average value for a χ^2 distribution with v d.o.f. is $v = (10-1)^2 = 81$ and the σ of the distribution is $\sqrt{2v} = \sqrt{2*81} = 12.73$ so we see that 591 is (591-81)/12.73 = 40 σ's above the average value of the distribution. We would thus conclude that a value of CC = 0.089 based upon 1000 data points using 10 equally populated bins is highly significant.

4.4 Rank Correlation

For many applications we are interested in locating variables among the candidate predictors that help us explain trends in the dependent variable. There are many examples of trends in most areas of science and technology. As examples, consider the following:

> **Agriculture**: The effect upon crop yield as a function of the amount of pesticides used.
> **Chemistry**: The relationship between reaction rate and the density of a particular species.
> **Econometrics**: The effect of interest rate upon the rate of inflation.
> **Medicine**: The relationship between cancer and long-term exposure to radiation.
> **Production Engineering**: The relationship between rejection rate and process time per part.

The simplest measure of a trend is the linear correlation coefficient ρ as described in Section 3.5. A trend can be either positive or negative and

ρ is limited to the range -1 to 1. Consider the 3 relationships for y as a function of x as shown in Figure 4.4.1. Curve A would yield a value of ρ between 0 and 1 and curve B would yield a value of ρ between 0 and -1. The value of ρ for curve C would be close to zero.

If our purpose is to identify trends, then we would like a measure that would yield a value of one for curve A because any increase in x results in an increase in y throughout the entire range of x. Similarly this measure should yield a value of -1 for curve B and a value close to zero for curve C. One might ask the question, why not just use the linear correlation coefficient ρ as our measure of a trend? When there are a number of candidate predictors for y it is useful to have a nonparametric measure that is more sensitive to trends than ρ. We use the term "nonparametric" because no specific relationship (like a straight line) is assumed between x and y. It thus becomes easier to find the candidate predictor (or predictors) that are most responsible for causing trends in y.

Probably the most well-known and widely used nonparametric measure of trends was proposed in 1904 by Spearman and is called the **Spearman Rank Correlation Coefficient** (i.e., r_s). Many books on statistics include a discussion of this measure of association between variables [e.g., SI88, DA90, FR92, ME92, WA93]. The method uses the ranks of the variables rather than the actual values. If several data points have the same value of either x or y, then the average rank is used.

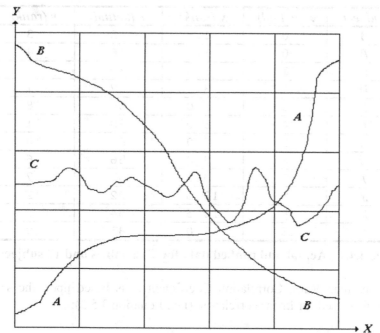

Figure 4.4.1   Three curves with differing trends

As an example, consider the data in Table 4.4.1. A rank of 1 is assigned to the lowest value of x and y and a value of n (the number of data points) is assigned to the highest value. The data in this table is based upon an experiment in sociology discussed by Siegel and Castellan [SI88]. For example, in this table the highest value of x is 13 and since there are 12 subjects it is assigned a rank of $x = 12$. The lowest value is 0 but since there are two subjects with $x = 0$ they are both assigned a rank of 1.5. The minimum and maximum values of y are 37 and 92 so the ranks of these two subjects are 1 and 12.

Subject	x (actual)	x (rank)	Y (actual)	y (rank)
A	0	1.5	42	3
B	0	1.5	46	4
C	1	3.5	39	2
D	1	3.5	37	1
E	3	5.0	65	8
F	4	6.0	88	11
G	5	7.0	86	10
H	6	8.0	56	6
I	7	9.0	62	7
J	8	10.5	92	12
K	8	10.5	54	5
L	13	12.0	81	9

Table 4.4.1   Actual and ranked data for 2 variables and 12 subjects.

The Spearman Rank Correlation Coefficient r_s is based upon the same equation as used for linear correlation (i.e., Equation 3.5.3):

$$r_s = \frac{\sum_{i=1}^{i=n}(x_i - x_{avg})(y_i - y_{avg})}{2\sqrt{\sum_{i=1}^{i=n}(x_i - x_{avg})^2 \sum_{i=1}^{i=n}(y_i - y_{avg})^2}} \tag{4.4.1}$$

Using the following equality (based upon the fact that $x_{avg} = y_{avg}$):

$$\sum_{i=1}^{i=n}(x_i - x_{avg})(y_i - y_{avg}) = \frac{\sum_{i=1}^{i=n}(x_i - x_{avg})^2 + \sum_{i=1}^{i=n}(y_i - y_{avg})^2 - \sum_{i=1}^{i=n}d_i^2}{2}$$

We obtain the following expression for r_s:

$$r_s = \frac{\sum_{i=1}^{i=n}(x_i - x_{avg})^2 + \sum_{i=1}^{i=n}(y_i - y_{avg})^2 - \sum_{i=1}^{i=n}d_i^2}{2\sqrt{\sum_{i=1}^{i=n}(x_i - x_{avg})^2 \sum_{i=1}^{i=n}(y_i - y_{avg})^2}} \tag{4.4.2}$$

In this equation, both x_{avg} and y_{avg} are $(n+1)/2$ and d_i is the difference between the x_i and y_i. The details required for the calculation of r_s are included in Table 4.4.2:

Subject	x (rank)	y (rank)	d	d^2	$(x-x_{avg})^2$	$(y-y_{avg})^2$
A	1.5	3	-1.5	2.25	25.00	12.25
B	1.5	4	-2.5	6.25	25.00	6.25
C	3.5	2	1.5	2.25	9.00	20.25
D	3.5	1	2.5	6.25	9.00	30.25
E	5.0	8	-3.0	9.00	2.25	2.25
F	6.0	11	-5.0	25.00	0.25	20.25
G	7.0	10	-3.0	9.00	0.25	12.25
H	8.0	6	2.0	4.00	2.25	0.25
I	9.0	7	2.0	4.00	6.25	0.25
J	10.5	12	-1.5	2.25	16.00	30.25
K	10.5	5	5.5	30.25	16.00	3.25
L	12.0	9	3.0	9.00	30.25	6.25
Total	78	78	0.0	109.5	141.5	143.0

Table 4.4.2   Actual and ranked data for 2 variables and 12 subjects.

For this example, the average value of the ranks of both x and y is 6.5. For subject A the value of d_i is 1.5–3 = -1.5, the value of x_i-x_{avg} is 1.5–6.5 = -5 and the value of y_i-y_{avg} is 3–6.5 = -3.5. The value of r_s for this example is:

$$r_s = \frac{141.5 + 143 - 109.5}{2\sqrt{141.5 * 143}} = \frac{175}{2 * 142.2} = 0.615$$

This seems like a fairly large correlation coefficient, but is it significant? Siegel includes critical values of r_s for testing significance for n up to 50 [Table Q in SI88]. For example, for n equal to 12, the critical value for a 2.5% level of confidence is 0.587 and for 1% it is 0.671. In other words, at a 2.5% level of confidence, 0.615 is significant, but it is not significant at a 1% level of confidence. If x and y are totally independent variables and we repeat the experiment many times, we would expect to compute a value of $r_s \geq 0.615$ between 1% and 2.5% of the time. For large values of n the parameter z is approximately normally distributed:

$$z = r_s\sqrt{n-1} \quad \text{(normally distributed for large } n) \qquad (4.4.3)$$

For example, for $n = 1000$, the critical value for a 1% level of confidence would require a z value = 2.326:

$$r_s = 2.326/\sqrt{999} = 0.0736 \quad \text{(critical value for 1% level of confidence)}$$

Thus using a 1% level of confidence, if $n = 12$ only values of r_s greater than 0.671 are considered significant but if n is increased to 1000, values greater than only 0.0736 are considered significant.

Even though Equation 4.4.3 is valid for large n can it be used for fairly small values of n? For example, from Siegel's table we know that the 1% confidence limit for $n = 12$ is 0.671. What value would we get using Eq 4.4.3? Surprisingly, the value is very close:

$$r_s = z/\sqrt{n-1} = 2.236/\sqrt{11} = 0.674 \quad \text{(for 1% confidence level)}$$

In Table 4.4.3 critical values of r_s for a 1% confidence level are listed for various values of n. A critical value of one implies that any value of r_s less than one is not significant.

n	r_s (Siegel table)	r_s (Equation 4.4.3)
5	1.000	>1.000
6	0.943	1.000
7	0.893	0.913
8	0.833	0.845
9	0.783	0.791
10	0.745	0.745
11	0.709	0.707
12	0.671	0.674
15	0.604	0.598
20	0.520	0.513
50	0.329	0.320

Table 4.4.3   Critical values of r_s for 1% confidence level.

Chapter 5 DESIGNING QUANTITATIVE EXPERIMENTS

5.1 Introduction

Designing a quantitative experiment implies choosing the number of data points, selecting the values of the independent variable (or variables), and when possible, setting the accuracy to which the individual data points are to be obtained. What is assumed is the mathematical model that is the basis of the proposed experiment. The design process predicts the accuracy of the results that the least squares process should yield prior to actually obtaining any data. By varying the experimental variables, the analyst can determine what has to be done so that the experiment should meet the proposed accuracy objectives. Alternatively, the analyst might conclude that the experiment, as proposed, will not succeed (i.e., meet the accuracy objectives).

As an example, consider an experiment to determine the half-life of a radioactive isotope. What is known is that the half-life is about 1 second, and the purpose of the experiment is to accurately measure this half-life to an accuracy of about 1%. This experiment was discussed in Section 1.1 and the proposed mathematical model is Equation 1.1.1. Let us recast this equation using simpler notation:

$$y = a_1 \cdot e^{-a_2 \cdot t} + a_3 \qquad (5.1.1)$$

The half-life is inversely proportional to a_2 (see Equation 1.1.2) so to meet the experimental requirements we would have to determine a_2 to 1% accuracy. The value of a_1 can be controlled to a certain extent by controlling the time the specimen is irradiated. But clearly there is a practical limit based upon the resources available to the experimenter. The background count-rate (i.e., a_3) can be estimated or even measured prior to actually running the experiment using the radioactive isotope.

The actual experiment will be performed by using apparatus that measures the number of counts in a time window of Δt seconds. This experiment is best understood if the units of y, a_1 and a_3 are all in *cps* (counts per second). To convert the Y's (the measured values of counts in the time windows) to *cps* we must divide by Δt (i.e., $y_i = Y_i / \Delta t$). Thus a_1 is the number of *cps* at time t equal to zero and a_3 is the background rate in units of *cps*. The number of data-points that can be obtained is limited because the count-rate decreases with time until it approaches the background count-rate. A more practical limit is to select a "reasonable" number of half-lives to run the experiment. For example, if the half-life is 1 second, it makes no sense to run the experiment for 100 seconds because the first term in Equation 5.1.1 is infinitesimal after 100 seconds. If the half life is about 1 second, then after 5 seconds this term has decreased to a value that is only about 3% of the value at time zero. In other words, after 5 half lives, the signal is reduced by a factor of $2^5 = 32$. Let us say we set the total time of the experiment to 5 seconds, thus the number of data points is $5 / \Delta t$. For counting experiments we know that the estimated standard deviation σ_Y of each value of Y is $sqrt(Y)$ so the relative uncertainties of the data points are:

$$\frac{\sigma_y}{y} = \frac{\sqrt{Y}}{Y} = \frac{\sqrt{y\Delta t}}{y\Delta t} = \frac{1}{\sqrt{\Delta t}\sqrt{y}} \qquad (5.1.2)$$

Note that the relative uncertainly in y is inversely proportional to the square root of Δt. Thus if Δt is halved, then the number of data points is doubled, but the relative uncertainty of each point is increased by a factor of $sqrt(2)$.

The experimentalist planning an experiment of this type would first estimate the background term a_3 and the expected value of a_1 (i.e., the actual signal). The maximum value of signal to noise is a_1 / a_3. The next task is to estimate the effects of a_1 / a_3, n and Δt upon the expected value of the ratio σ_{a2} / a_2. The target value for this ratio is 0.01. At the design level we should be able to either choose the experimental variables so that this target can be achieved or conclude that the target is unattainable using the available resources. In this chapter the methodology for estimating expected accuracy is developed. The method of Prediction Analysis [WO67] is described and examples of its application are included. A more detailed analysis of this experiment is included in Section 5.3.

5.2 The Expected Value of the Sum-of-Squares

The method of least squares is based upon the minimization of S, the weighted sum of the squares of the differences between actual and computed values of the dependent variable. In Section 2.2 several different formulations of S are presented. In Section 3.2 the goodness-of-fit of a least squares model is discussed and it is shown that under certain conditions the expected value of S is χ^2 (chi-squared) distributed with $n - p$ degrees of freedom. This fact can be exploited to design experiments.

Under the assumption that the weights w_i are based upon reasonable estimates of the standard deviations of the residuals R_i (i.e., the difference between actual and computed values), the expected value of S is $n - p$ and S/n–p is one. Substituting this value into Equation 2.5.1 we see that the predicted value of the standard deviation of a_k is:

$$\sigma_{a_k} = (C_{kk}^{-1})^{1/2} \tag{5.2.1}$$

In this equation C_{kk}^{-1} is the predicted value of the kth diagonal term of the inverse C matrix. From Equation 2.5.4, the predicted value of the covariance between the parameters a_j and a_k is:

$$\sigma_{jk} = C_{jk}^{-1} \tag{5.2.2}$$

We thus see that the design of experiments is based upon prediction of the C matrix. The terms of the C matrix are computed using Equation 2.4.14. Once the terms of the C matrix have been computed, the matrix can then be inverted and all variances associated with the results of the experiment can be predicted. By varying the experimental variables, we can study the effect that these variables will have upon the resulting accuracy of the proposed experiment. An understanding of these effects is the basis of a rational design of the experiment.

5.3 The Method of Prediction Analysis

Prediction Analysis is the name that I used to describe a technique that I developed for designing quantitative experiments [WO67]. As explained in the previous section, the method of prediction analysis requires prediction of the C matrix. The terms of this matrix are computed as described in Section 2.4 by Equation 2.4.14:

$$C_{jk} = \sum_{i=1}^{i=n} w_i \frac{\partial f}{\partial a_j} \frac{\partial f}{\partial a_k} \qquad (2.4.14)$$

To demonstrate the method, let us consider the experiment discussed in Section 5.1: measurement of the half-life of a radioactive isotope in the presence of some background radiation. The function f for this experiment includes a decaying exponential term:

$$y = f(t) = a_1 \cdot e^{-a_2 \cdot t} + a_3 \qquad (5.3.1)$$

The experimental variables for this particular class of experiments are the three unknown parameters, a_1, a_2 and a_3, the times of the initial and final measurements t_0 and t_{max}, the number of data points n and the time window Δt. The three partial derivatives are:

$$\frac{\partial f}{\partial a_1} = e^{-a_2 \cdot t} \qquad (5.3.2)$$

$$\frac{\partial f}{\partial a_2} = -a_1 t e^{-a_2 \cdot t} \qquad (5.3.3)$$

$$\frac{\partial f}{\partial a_3} = 1 \qquad (5.3.4)$$

The weights w_i are computed as follows using Equation 5.1.2:

$$w_i = 1 / \sigma_{y_i}^2 = \Delta t / y_i \qquad (5.3.5)$$

The terms of the C matrix are computed according to Equation 2.4.14. Using Equation 5.3.5:

$$C_{jk} = \sum_{i=1}^{i=n} w_i \frac{\partial f}{\partial a_j} \frac{\partial f}{\partial a_k} = \Delta t \sum_{i=1}^{i=n} \frac{1}{y_i} \frac{\partial f}{\partial a_j} \frac{\partial f}{\partial a_k} \qquad (5.3.6)$$

Equation 5.3.6 is straightforward. All that one must do is select a set of the experimental variables and then compute the terms of the C matrix by substituting Equation 5.3.1 through 5.3.4 into 5.3.6. The matrix is symmetric so only 6 terms must be computed: C_{11}, C_{12}, C_{13}, C_{22}, C_{23} and C_{33}. Once these terms have been computed the matrix is inverted and then Equations 5.2.1 and 5.2.2 can be used to make predictions regarding the expected accuracy of the proposed experiment.

The seven variables mentioned above are a_1, a_2, a_3, t_0, t_{max}, n and Δt. If we assume that $t_0 = 0$ and $t_{max} = n*\Delta t$ we are left with 5 independent variables. There are two approaches that one might consider before initiation of the calculations:

1) Try to develop analytical expressions for the 6 terms of the C matrix, and if successful, then try to develop analytical expressions for the terms of the C^{-1} matrix.

2) Use computer simulations to compute the terms of the C and C^{-1} matrices for various combinations of the variables.

The first approach is only feasible for very simple experiments. An example of this approach is considered in Sections 5.4 and 5.5. The class of experiments discussed in this section is, unfortunately, too complicated to attempt an analytical approach. Equations can be developed, but they are too cumbersome to be useful. Results, however, can easily be obtained using the second approach: computer simulations. A computer simulation for this class of experiments is discussed in Section 5.6.

To understand the complexity of trying to develop analytical expressions for the terms of the C matrix, consider just the term C_{11} :

$$C_{11} = \Delta t \sum_{i=1}^{i=n} \frac{1}{y_i} \frac{\partial f}{\partial a_1} \frac{\partial f}{\partial a_1} = \Delta t \sum_{i=1}^{i=n} \frac{1}{y_i} e^{-2a_2 t_i}$$

$$= \frac{n\Delta t}{a_1} \left[\frac{e^{-2a_2 t}}{e^{-a_2 t} + a_3/a_1} \right]_{avg}$$

(5.3.7)

We can get an estimate of the average value in Equation 5.3.7 for large n by integrating from 0 to $n\Delta t$ and then dividing the result by $n\Delta t$:

$$\left[\frac{e^{-2a_2 t}}{e^{-a_2 t} + a_3/a_1} \right]_{avg} \cong \frac{1}{n\Delta t} \int_0^{n\Delta t} \frac{e^{-2a_2 t}}{e^{-a_2 t} + a_3/a_1} \, dt$$

(5.3.8)

Using $T = n\Delta t$ and $b = a_3/a_1$, the integral in Equation 5.3.8 is

$$\int_0^T \frac{e^{-2a_2 t}}{e^{-a_2 t} + b} \, dt = \frac{1}{a_2} \left(1 - e^{-a_2 T} - \frac{a_3}{a_1} \log \left(\frac{1+b}{e^{-a_2 T} + b} \right) \right)$$

Although we can get an analytical expression for C_{11} it is complicated and when combined with similar expressions for the other terms of the C matrix, we are left with a matrix that term by term can be estimated analytically but when we invert this matrix, the analytical expressions are too complicated to be useful.

An important aspect of the *prediction analysis* is presentation of the results. For the class of experiments discussed in this section we reduced our initial set of 7 variables to 5 by assuming that the experiments start at time $t = 0$ and that t_{max} is equal to $n\Delta t$. We can do a lot better than 5 variables if we combine the variables into dimensionless groups. Two obvious dimensionless groups that characterize this experiment are the dimensionless background a_3/a_1 and the dimensionless duration of the experiment z (i.e., $a_2 * t_{max}$). In Section 5.6 results are presented graphically as functions of only these two variables.

5.4 A Simple Example: A Straight Line Experiment

Probably the most frequently used experimental model is a straight line in which x is the independent variable and y is the dependent variable:

$$y = a_1 + x \cdot a_2 \qquad (5.4.1)$$

The purpose of the experiment might be computation of the parameter a_1 and/or a_2 or perhaps the purpose might be determination of an equation that can be used for interpolation. Whatever the purpose, to design the experiment it is important to understand how the expected accuracy of the results should be dependent upon the experimental variables.

The starting points for an analysis are models for the expected uncertainties associated with the measured values of x and y (i.e., σ_{x_i} and σ_{y_i}). Let us assume that both variables can be characterized as having constant uncertainty:

$$\sigma_{x_i} = K_x$$
$$\sigma_{y_i} = K_y \qquad (5.4.2)$$

From Equation 2.3.7 we see that the weights for all points are the same:

$$w_i = \frac{1}{(\sigma_{y_i}^2 + (\frac{\partial f}{\partial x}\sigma_{x_i})^2)} = \frac{1}{(K_y^2 + (a_2 K_x)^2)} = W \qquad (5.4.3)$$

Examining Equations 2.4.10 and 2.4.11 we see that the terms of the C matrix can easily be determined:

$$C = \begin{bmatrix} \sum\limits_{i=1}^{i=n} w_i & \sum\limits_{i=1}^{i=n} w_i x_i \\ \sum\limits_{i=1}^{i=n} w_i x_i & \sum\limits_{i=1}^{i=n} w_i x_i^2 \end{bmatrix} = \begin{bmatrix} nW & W\sum\limits_{i=1}^{i=n} x_i \\ W\sum\limits_{i=1}^{i=n} x_i & W\sum\limits_{i=1}^{i=n} x_i^2 \end{bmatrix}$$

$$= nW \begin{bmatrix} 1 & x_{avg} \\ x_{avg} & (x^2)_{avg} \end{bmatrix} \qquad (5.4.4)$$

We are now in a position to estimate the values of σ_{a_1} and σ_{a_2} that we can expect from the straight-line experiment using Equation 5.2.1:

$$\sigma_{a_1} = (C_{11}^{-1})^{1/2} = \left(\frac{(x^2)_{avg} / nW}{(x^2)_{avg} - (x_{avg})^2} \right)^{1/2} \tag{5.4.5}$$

$$\sigma_{a_2} = (C_{22}^{-1})^{1/2} = \left(\frac{1/nW}{(x^2)_{avg} - (x_{avg})^2} \right)^{1/2} \tag{5.4.6}$$

All that remains is to choose values of x_i and then determine the average values of x and x^2. For example, assume that the experiment will be performed using values of x equally spaced from x_1 to x_n.

$$\Delta x = x_{i+1} - x_i = (x_n - x_1)/(n-1) \tag{5.4.7}$$

The average value of x is simply $(x_1 + x_n)/2$. The average value of x^2 can be estimated using the following equation:

$$
\begin{aligned}
(x^2)_{avg} &= \int_{x_1}^{x_n} x^2 dx / (x_n - x_1) \\
&= \frac{1}{3} \frac{(x_n^3 - x_1^3)}{(x_n - x_1)}
\end{aligned}
\qquad \text{as } n \rightarrow \infty \tag{5.4.8}
$$

For small values of n the accuracy of this estimation can be improved by using $x_1 - \Delta x/2$ and $x_n + \Delta x/2$ as the limits of integration:

$$
\begin{aligned}
x_a &= x_1 - \Delta x / 2 \\
x_b &= x_n + \Delta x / 2
\end{aligned}
\tag{5.4.9}
$$

Using these limits of integration we get the following:

$$(x^2)_{avg} = \int_{x_a}^{x_b} x^2 dx / (x_b - x_a)$$

$$= \frac{1}{3} \frac{(x_b^3 - x_a^3)}{(x_b - x_a)} = \frac{1}{3}(x_b^2 + x_a x_b + x_a^2) \tag{5.4.10}$$

Substituting 5.4.10 into 5.4.5 and 5.4.6, after some algebraic manipulations [WO67] we get the following estimations of the standard deviations that can be expected from a straight line experiment with equally spaced points and constant values of σ_{x_i} and σ_{y_i}:

$$\sigma_{a_1} = \left(\frac{4}{nW} \frac{x_b^3 - x_a^3}{(x_b - x_a)^3} \right)^{1/2} \tag{5.4.11}$$

$$\sigma_{a_2} = \left(\frac{12}{nW} \frac{1}{(x_b - x_a)^2} \right)^{1/2} \tag{5.4.12}$$

Noting that $x_b - x_a = n\Delta x$, we can put Equation 5.4.11 into a physically meaningful form using the following dimensionless parameter:

$$r_x = \frac{x_{avg}}{x_b - x_a} = \frac{x_{avg}}{n\Delta x} \tag{5.4.13}$$

This parameter is the midpoint of the x values normalized by the range of the x values. It can be shown that Equation 5.4.11 is equivalent to the following [WO67]:

$$\sigma_{a_1} = \left(\frac{12r_x^2 + 1}{nW} \right)^{1/2} \tag{5.4.14}$$

We thus see that if the purpose of the experiment is to measure a_1, the best design is to set $r_x = 0$ by centering the points about $x = 0$. Clearly this is not always possible. For example, if only positive values of x are possible, then the best that we could do is set $r_x = 1/2$.

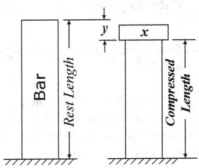

Figure 5.4.1   An Experiment to Measure the Coefficient of Elasticity of a Bar

The purpose of the preceding analysis is to develop equations that can be used for the design of an experiment. To demonstrate how one would use the equations, let us design an experiment to measure the coefficient of elasticity of a bar in compression as shown in Figure 5.4.1. The independent variable x is the weight placed upon the bar (in newtons) and the dependent variable y is the measured compression (in cms). Within the elastic range, the behavior is assumed to be linear (i.e., Equation 5.4.1) and the elastic coefficient is the slope of the line (i.e., a_2 with units cm/n). Let us assume that the uncertainty of the weights are negligible (i.e., $K_x = 0$) and the uncertainty of the measured values of y are 0.1 (i.e., $K_y = 0.1$). Using Equation 5.4.3) the value of W is therefore 100. Let us assume that we wish to measure a_2 to an accuracy of 0.01 cm/n and that the range of weights that is feasible for this experiment is 10 newtons. How many different weights are needed to achieve the desired accuracy? Inserting these numbers into Equation 5.4.12 we get the following:

$$\sigma_{a_2} = \left(\frac{12}{nW} \frac{1}{(x_b - x_a)^2} \right)^{1/2} \tag{5.4.15}$$

$$= \left(\frac{12}{100n} \frac{1}{10^2} \right)^{1/2} = \frac{3.464}{100n^{1/2}} = 0.01$$

Solving 5.4.15 for n we get a value of $n = 12$. Note that this value is not a function of r_x. Thus the resulting slope could theoretically be measured from 0 to 10 newtons or from 5 to 15 newtons and the resulting accuracy of the measured value of a_2 could be expected to be about the same. To improve the accuracy we have 3 possibilities:

1) Improve the accuracy of measured values of y (i.e., reduce K_y).
2) Extend the range (i.e., increase $x_b - x_a$).
3) Increase n.

A byproduct of the experiment is the resulting value of a_1 and we would expect this value to be close to zero. If this value is significantly different from zero we would have to question the applicability of Equation 5.4.1. Let us assume that the experiment uses a range starting close to $x = 0$ and therefore r_x would be 1/2. From Equation 5.4.12 we would expect a value of $\sigma_{a_1} = 0.058$. If after the experiment has been performed, if the least squares analysis of the data yields a value of a_1 that is outside a reasonable range (for example a 2σ range: $-0.058 * 2 < a_1 < 0.058 * 2$), then one should seriously question all aspects of the experiment including the applicability of the mathematical model (i.e., Equation 5.4.1).

5.5 Designing for Interpolation

The purpose of some experiments is to create a function that can be used for interpolation. For such experiments the values of the unknown parameters of the function are not of particular interest. They are, however, of interest when taken together to be used for interpolation for any combination of the independent variable (or variables) within the range of the experiment. To illustrate how one would go about designing such an experiment, we can continue the analysis started in the previous section: the straight line. In this section, the analysis is directed towards the resulting line that is to be used for interpolation.

Using Equation 5.4.1 and the constant weight assumption (Eqs. 5.4.2 and 5.4.3), we start our analysis from Equation 5.4.4: the expressions for the terms of the C matrix. This matrix can be inverted to yield the terms of the C^{-1} matrix:

$$C^{-1} = \frac{1}{Det} \begin{bmatrix} \sum_{i=1}^{i=n} w_i x_i^2 & -\sum_{i=1}^{i=n} w_i x_i \\ -\sum_{i=1}^{i=n} w_i x_i & \sum_{i=1}^{i=n} w_i \end{bmatrix}$$

$$= \frac{nW}{Det} \begin{bmatrix} (x^2)_{avg} & -x_{avg} \\ -x_{avg} & 1 \end{bmatrix} \tag{5.5.1}$$

where **Det** is the determinant of the **C** matrix:

$$Det = \sum_{i=1}^{i=n} w_i x_i^2 \sum_{i=1}^{i=n} w_i - (\sum_{i=1}^{i=n} w_i x_i)^2$$

$$= n^2 W^2 ((x^2)_{avg} - (x_{avg})^2) \tag{5.5.2}$$

Substituting 5.5.2 into 5.5.1 we get the following expression for the inverse **C** matrix:

$$C^{-1} = \frac{1}{nW((x^2)_{avg} - (x_{avg})^2)} \begin{bmatrix} (x^2)_{avg} & -x_{avg} \\ -x_{avg} & 1 \end{bmatrix} \tag{5.5.3}$$

We can now proceed to predict the σ's associated with interpolations that should result from a straight line experiment. We use Equation 2.6.11 as our starting point:

$$\sigma_f^2 = \frac{S}{n-p} \sum_{j=1}^{j=p} \sum_{k=1}^{k=p} \frac{\partial f}{\partial a_j} \frac{\partial f}{\partial a_k} C_{jk}^{-1} \tag{2.6.11}$$

Noting that the expected value of $S / (n - p)$ is 1, the predicted σ's are computed using the modified form of 2.6.11:

$$\sigma_f^2 = \sum_{j=1}^{j=p} \sum_{k=1}^{k=p} \frac{\partial f}{\partial a_j} \frac{\partial f}{\partial a_k} C_{jk}^{-1} \tag{5.5.4}$$

For the straight line experiment we can carry this equation through to an analytical solution, but for most relationships between x and y the best that we can do is to simulate the experiment. For the straight line, the partial derivatives are:

$$\frac{\partial f}{\partial a_1} = 1 \qquad \text{and} \qquad \frac{\partial f}{\partial a_2} = x \qquad (5.5.5)$$

Substituting 5.5.5 into 5.5.4 and noting that $C_{12}^{-1} = C_{21}^{-1}$ we get the following:

$$\sigma_f^2 = C_{11}^{-1} + 2xC_{12}^{-1} + x^2 C_{22}^{-1} \qquad (5.5.6)$$

Using the equations developed in Section 5.4 for the average values of x and x^2 and then substituting them into Equation 5.5.3 and then 5.5.6, we get the following equation for the variance that can be expected for any value of x:

$$\sigma_f^2 = \frac{4}{nW(x_b - x_a)^2}(x_b^2 + x_a x_b + x_a^2 - 3x(x_b + x_a) + 3x^2) \quad (5.5.7)$$

This equation is parabolic with minimum variance at the midpoint of the range (i.e., $x = (x_a + x_b) / 2$). Substituting this value of x into 5.5.7 we get the expected minimum value of the variance:

$$(\sigma_f^2)_{\min} = \frac{4}{nW(x_b - x_a)^2}(\frac{1}{4}(x_b^2 + x_a^2) - \frac{1}{2}x_a x_b) \qquad (5.5.8)$$

The expected maximum variance is noted at the extremes of the range (i.e., $x = x_a$ and $x = x_b$):

$$(\sigma_f^2)_{\max} = \frac{4}{nW(x_b - x_a)^2}(x_b^2 - 2x_a x_b + x_a^2) \qquad (5.5.9)$$

To design an experiment we should establish the maximum variance as a design objective. For example, let us assume that the measurements of y will be accurate to 0.1 (i.e., $K_y = 0.1$) and therefore $W = 100$ (i.e., $1 / K_y^2$).

Let us further assume that the data points will be equally spaced from $x =$ 0.1 to 0.3. Setting a design objective of maximum standard deviation = 0.04 (i.e., $\sigma_f \leq 0.04$), we can use equation 5.5.9 to determine the number of points needed to theoretically meet this objective (at least in the design of the experiment):

$$
(\sigma_f^2)_{max} = 0.04^2 = \frac{4/100}{n*0.2^2}(0.3^2 - 2*0.1*0.3 + 0.1^2)
$$

$$
= \frac{0.04*0.04}{0.04*n} = \frac{0.04}{n}
$$

(5.5.9)

Solving for n we get a value of $n = 1 / 0.04 = 25$. In other words, if we equally space 25 point in the range x from 0.1 to 0.3, we should obtain a line that if used for interpolation will yield values of y with σ_f no worse than 0.04. It should be remembered that for the purpose of designing the experiment we have assumed a value of $S / (n - p) = 1$, so when the experimental data is actually analyzed, the values of σ_f might be larger or smaller than the design values. Nevertheless, the design process helps us choose the experimental variables that should yield results that are reasonably close to our design objectives.

In this section the mathematical model was simple enough to allow us to develop analytical equations that were useful in designing for interpolation. When the mathematical model is complicated, the use of computer simulations is the obvious approach to design. In the following section a more complicated experiment is analyzed and predicted values for interpolations at 10 points are included in a simulation (i.e., Figure 5.6.2). In the figure the column headed PRED-SIGY are the predicted values of σ_f.

5.6 Design Using Computer Simulations

In Section 5.3 we discussed an experiment to measure half-life of a radioactive isotope based upon the following exponentially decaying mathematical model:

$$
y = f(t) = a_1 \cdot e^{-a_2 \cdot t} + a_3
$$

(5.3.1)

The experiment is performed by observing the number of counts recorded in n time windows of length Δt. We noted that two dimensionless groups that characterize this experiment are the dimensionless background a_3 / a_1 and the dimensionless duration of the experiment z (i.e., $a_2 t_{max} = a_2 n\Delta t$).

In Section 5.4 we developed analytical expressions for the unknown parameters in a simple straight-line experiment but for experiments based upon Equation 5.3.1, the complexity of the equations led to the conclusion that for these experiments, the most reasonable approach is to use computer simulations. This conclusion was reached after examining the equations needed to compute terms of the C matrix (e.g., Equation 5.3.7 thru 5.3.9). Not only is this a 3 by 3 matrix, it must also be inverted to obtain C_{22}^{-1}. This term is needed because the purpose of the experiment is to measure a_2 including an estimate of its standard deviation. The predicted value of this standard deviation is :

$$\sigma_{a2}^2 = C_{22}^{-1} \qquad (5.6.1)$$

To perform the simulation, it should be realized that the n values of the t_i's should be set at the middle of the time windows rather than at the beginning or the end. If n is large then this effect is negligible, but for small n the effect can be noticeable. Thus if the experiment is started at $t = 0$, the value of $t_1 = \Delta t/2$ and $t_n = t_{max} - \Delta t/2$. Results from a series of simulations confirm that the following dimensionless group Ψ_2 is only a function of a_3/a_1 and z:

$$\Psi_2 = \frac{\sigma_{a2}}{a_2}\left(a_1 n\Delta t\right)^{1/2} = \frac{\sigma_{a2}}{a_2}\left(a_1 t_{max}\right)^{1/2} = F(z, a_3/a_1) \qquad (5.6.2)$$

In Figure 5.6.1 Ψ_2 is plotted as a function of z for several different values of a_3 / a_1. The results in the figure were based upon simulations. An example of a simulation of one combination of the experimental parameters is shown in Figure 5.6.2. This simulation is for the combination $a_1 = 10000$, $a_2 = 1$, $a_3 = 500$, $z = 10$ and $n = 10$. The simulation was generated using the Prediction Analysis feature of the REGRESS program [see Section 6.8]. In the figure PRED_SA(K) is "predicted σ_{ak}" and PRED_SIGY is "predicted σ_f". Note that for this particular combination of a_2, z and n the value of Δt is one and t_{max} is 10. The 10 values of T and Y were generated using $\Delta t = 1$ and the input equation for Y starting from the mid-point of the first time window (i.e., $t_1 = 0.5$). Note that the predicted value of σ_{a2}/a_2 is 0.01799. Using Equation 5.6.2 the value of Ψ_2 is 0.01799 * $(10000 * 10)^{1/2} = 5.69$. Comparing this to the value of 5.107 listed in Table 5.6.1 we see that there is a difference of about 10%. This difference is due to the fact that the values in the table (and in Figure 5.6.1) were generated using a large value of n and can thus be considered as the asymptotic values. It is reassuring to see that using Figure 5.6.1 yields reasonable estimates of the predicted value of σ_{a2}/a_2 even for n as small as 10.

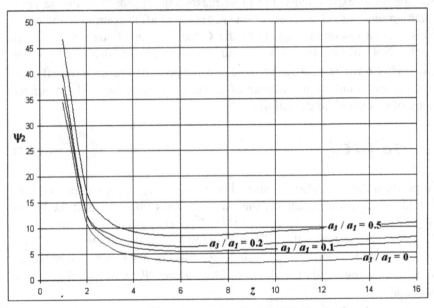

Figure 5.6.1 Ψ_2 versus z for Several Values of a_3 / a_1

These results can be used to predict the value of σ_{a2}/a_2 that can be expected from any combination of the experimental parameters. Using Figure 5.6.1 and Equation 5.6.2, the value of σ_{a2}/a_2 can be estimated for a range of combinations of 3 dimensionless parameters: a_3/a_1, z, and $a_1 t_{max}$. It should be noted that our objective was to measure half-life but Equation 5.6.2 shows results for the relative uncertainty that we should expect for a_2. Since half-life is inversely proportional to the decay constant a_2 (Equation 1.1.2), the relative uncertainty in the half-life h is the same as the relative uncertainty in a_2 :

$$\frac{\sigma_h}{h} = \frac{\sigma_{a2}}{a_2}$$

$(5.6.3)$

As an example of the use of Equation 5.6.2, consider the experimental design problem posed in Section 5.1: the measurement of a_2 to 1% accuracy for a radioactive isotope with a half-life of approximately 1 second. We see from Figure 5.6.1 that Ψ_2 is close to its minimum value at $z = 10$ so let us choose this value of z as the design value. Since a_2 is approximately equal to 0.693 / *half-life*, $a_2 = 0.693 / 1.0$, and $t_{max} = z / a_2 = 10 / 0.693 = 14.4$ seconds. In Table 5.6.1 values of Ψ_2 are listed for different values of the ratio a_3 / a_1 and then the value of a_1 required to obtain $\sigma_{a2} / a_2 = 0.01$ are computed. If the value of a_3 is estimated by a separate experiment, we can thus estimate the value of a_1 that yields the required accuracy. It should be remembered that we are not guaranteed that the resulting accuracy will be exactly 1%. The actual results of experiments based upon this design should be χ^2 distributed about this average value of 1%.

From the table we see that if the value of the noise ratio a_3 / a_1 is close to zero, then the design value of a_1 is about 7500 counts per second. As the noise ratio increases, the design value of a_1 must be increased. Let us say that we are limited to a value of a_1 equal to 15000. We would thus be limited to a noise ratio of about 0.04. In other words, the value of a_3 would have to be limited to about 0.04 * 15000 = 600 counts per second. Whether or not this is achievable is dependent upon the actual experimental equipment and environment. However, it is extremely useful to know this prior to actually running the experiment.

```
PARAMETERS USED IN REGRESS ANALYSIS: Thu Jun 24
08:58:27 2004
  INPUT PARMS FILE: tab561.par
  INPUT DATA  FILE: tab561.par
  REGRESS  VERSION: 4.08, Dec 22, 2003

  Prediction Analysis Option  (MODE='P')
  N - Number of recs used to build model  :     10
  NCOL - Number of data columns           :      0
  NY   - Number of dependent variables    :      1
  Y VALUES - computed using A0 & T values
  SYTYPE -    SIGMA Y = CY1 * sqrt(Y)    CY1: 1.000
  X Values - computed using interpolation table
  STTYPE - Sigma type for T               :      0

  Function Y:  A1*EXP(-A2*T) + A3
   K       A0(K)         A(K)    PRED_SA(K)
   1    10000.00     10000.00    186.04627
   2     1.00000      1.00000      0.01799
   3   500.00000    500.00000     10.03220


POINT            T        YCALC      PRED_SIGY
    1      0.50000   6565.30660      78.44070
    2      1.50000   2731.30160      33.92022
    3      2.50000   1320.84999      23.05382
    4      3.50000    801.97383      13.06935
    5      4.50000    611.08997       8.96969
    6      5.50000    540.86771       8.81757
    7      6.50000    515.03439       9.35975
    8      7.50000    505.53084       9.71965
    9      8.50000    502.03468       9.89662
   10      9.50000    500.74852       9.97538
```

Figure 5.6.2: Prediction Analysis for $z = 10$ and $a_3/a_1 = 0.05$.

a_3/a_1	ψ_2	$a_1 = (\psi_2/0.01)^2/t_{max}$
0.00	3.292	7526
0.05	5.107	18112
0.10	5.856	23814
0.20	6.941	33457
0.30	7.792	42163
0.40	8.526	50481
0.50	9.183	58564

Table 5.6.1 a_1 as function of a_3/a_1 for $z = a_2 t_{max} = 10$

5.7 Designs for Some Classical Experiments

In the book *Prediction Analysis* [WO67] I analyzed a number of classical experiments and included equations and graphs that allow the user to predict the σ_{a_k}'s for combinations of the experimental variables. The mathematical models considered in that book included polynomial functions (Chapter 5), exponential functions (Chapter 6), sine series (Chapter 7) and Gaussian functions (Chapter 8). Three alternative uncertainty models were analyzed for each mathematical model:

1) Constant uncertainty: $\sigma_{y_i} = K_y$

2) Constant fractional uncertainty: $\sigma_{y_i} = K_y y_i$

3) Counting statistical uncertainty: $\sigma_{y_i} = K_y y_i^{1/2}$

The uncertainties for the independent x variable were assumed to be either 0 or a constant value K_x. In this chapter two different cases were discussed in the previous section:

1) The first order polynomial function (i.e., the straight line) with constant uncertainty in the values of x_i and y_i (Section 5.4 and 5.5).

2) The exponential function with background (Equation 5.3.1) and with counting statistics as the uncertainty model for the values of y_i (Sections 5.3 and 5.6).

The first case was chosen to illustrate an analytical approach to design in which equations used to predict the σ_{a_k}'s were developed. The second case was chosen to illustrate the use of simulations to develop graphs for predicting the σ_{a_k}'s when development of equations is too cumbersome.

In this section three additional models are considered. The choice of models is based upon their usefulness and includes the following:

1) **Experiment 1:** A straight-line model with equally spaced x values and constant fractional uncertainty in the values of y.

2) **Experiment 2:** A decaying exponential model with equally spaced t (time) values and uncertainty in the values of y based upon counting

statistics. For this experiment the background term is assumed to be negligible.

3)   **Experiment 3:** A Gaussian peak model with equally spaced x values and uncertainty in the values of y based upon counting statistics.

Experiment 1 :

A useful variation of the model considered in Sections 5.4 and 5.5 is the straight line but with constant fractional uncertainty for the data points. It is assumed that the n data points will be equally spaced along the x axis and all values of x are greater or equal to zero. The simple equations developed for constant uncertainty can be used when the range of the y values is not large. However, when there are considerable differences in the values of y (for example, y_n / y_1 is either much greater or much less than 1, the simple equations are not applicable. Figure 5.7.1 illustrates the type of data that one might expect in such experiments. A common usage for such experiments is to fit a line to the data in order to determine the value of a_1 (i.e., y for $x = 0$) when it is impossible to measure the value directly. Also, many experiments are performed in which the purpose is to determine the slope of the line.

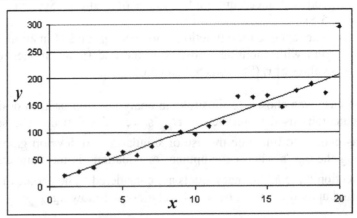

Figure 5.7.1   Straight Line Fit to Data with Constant Fractional Error

Analysis of this class of experiments yields equations that are extremely complicated and are therefore not particularly useful. Results are presented graphically in Figures 5.7.2 and 5.7.3 for two dimensionless groups:

$$\theta_1 = \frac{\sigma_{a1} n^{1/2}}{K_y y_{max}}$$ (5.7.1)

$$\theta_2 = \frac{\sigma_{a2}(x_n - x_1) n^{1/2}}{K_y y_{max}}$$ (5.7.2)

The denominator of these groups is the value of σ_y for the data point with the greatest uncertainty (either $K_y y_1$ or $K_y y_n$). The results for these groups are presented as functions of y_n/y_1. The results are asymptotic values as n approaches infinity but are good approximations even for small values of n. The first group is also a function of r_x (see Equation 5.4.13). Note that the results are exactly the same as computed using Equations 5.4.14 and 5.4.15 when $y_n/y_1 = 1$ if we use $K_y y_{max}$ in place of K_y in these equations.

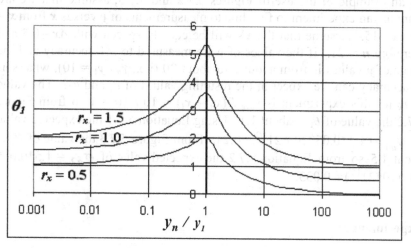

Figure 5.7.2 θ_1 versus y_n/y_1 for several values of r_x

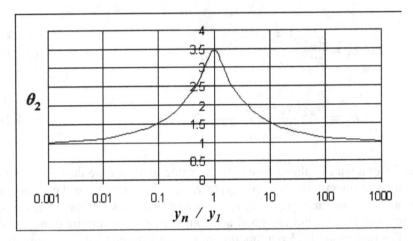

Figure 5.7.3 θ_2 versus y_n / y_1

As an example of the use of Figures 5.7.2 and 5.7.3, assume that we are planning an experiment to fit a line to measurements of y versus x from $x =$ 5 to $x =15$. Assume that the x's will be equally spaced with $\Delta x =0.5$ and therefore $n = 21$. If the values of y are measured to 5% accuracy and the range of y values is from about 2 to about 20 (i.e., $y_n / y_1 = 10$), what sort of accuracy can we expect in the resulting values of a_1 and a_2? The value of r_x for this experiment is $x_{avg} / (x_n - x_1) = 10 / 10 = 1$ so from Figure 5.7.2 the value of θ_1 is about 1.1. Using Equation 5.7.1 the expected value of $\sigma_{a1} = 1.1*0.05*20 / 21^{1/2} = 0.24$. From Figure 5.7.3 the value of θ_2 is about 1.5 so from Equation 5.7.2 the expected value of $\sigma_{a2} = 1.5*0.05*$ $20 /(10*21^{1/2}) = 0.033$.

Experiment 2:

Another useful experiment is similar to the experiment described in Section 5.3 (i.e., measuring the count rate of a radioactive isotope as a function of time) but without a significant background count rate. The uncertainties for the values of y are assumed to be based upon counting statistics. In place of Equation 5.3.1, the following mathematical model is used:

$$y = f(t) = a_1 \cdot e^{-a_2 \cdot t} \qquad (5.7.3)$$

Since there is no background count rate, we can define a_1 as the number of counts per Δt seconds and therefore the uncertainties in the measured values of y will be $y_i^{1/2}$. Defining z as in Section 5.6 (i.e., $z = a_2\, t_{max} = a_2 n \Delta t$), an equation can be developed for estimating σ_{a2} [WO67]:

$$\Psi_2 = \frac{\sigma_{a2}}{a_2}(a_1 n)^{1/2} = \left[\frac{z(1 - e^{-z})}{Det}\right]^{1/2} \qquad (5.7.4)$$

Where the determinant of the C matrix is computed as follows:

$$Det = (1 - e^{-z})\,(2 - e^{-z}(z^2 + 2z + 2)) - (1 - e^{-z}(z + 1))^2 \qquad (5.7.5)$$

As an example of the usage of Equation 5.7.4, consider an experiment in which a_2 is expected to be approximately 2 (sec^{-1}) but we wish to measure it to 1% accuracy. Let us choose a design value of $z = 10$ based upon $n = 100$ and $\Delta t = 0.05$ sec. (Note: $z = a_2 n \Delta t$). Substituting this number into Equation 5.7.5 we see that Det is very close to 1 and therefore Ψ_2 is very close to $10^{1/2}$. Solving for a_1:

$$a_1 = \frac{\Psi_2^2}{n(\sigma_{a2}/a_2)^2} = \frac{10}{100(0.01)^2} = 1{,}000 \qquad (5.7.6)$$

Is there an optimum value of z for running this experiment? The optimum value is the value that minimizes a_1. Values of Ψ_2 are shown in Table 5.7.1 for various values of z. We see that Ψ_2 is minimized at about $z = 4$. For this value of z we only need about 585 counts in the first time window to achieve an expected accuracy of 1% in the measured value of a_2.

z	Ψ_2	a_1
1.0	4.465	1994
2.0	2.895	838
3.0	2.503	627
3.5	2.439	595
4.0	2.419	585
4.5	2.431	591
5.0	2.463	607
10.0	3.169	1004
15.0	3.871	1498
20.0	4.447	1978

Table 5.7.1 Ψ_2 and a_1 versus z. a_1 is computed using Equation 5.7.6.

Experiment 3:

The final experiment discussed in this section is based upon data that is modeled using a Gaussian peak with unknown height (a_1), peak width (a_2), and peak location (a_3). The mathematical model for the values of y as a function of x is:

$$y = a_1 e^{-\left(\frac{x-a_3}{a_2}\right)^2}$$

(5.7.7)

Typically for experiments that are based upon this model data is recorded in some sort of multi-channel analyzer for some fixed time duration of the experiment. The x's are the mid-points of the channels. At the end of the experiment the number of counts recorded in each channel is analyzed based upon Equation 5.7.7. The uncertainties for the values of y are assumed to be based upon counting statistics: $\sigma_{y_i} = y_i^{1/2}$. In Figure 5.7.4, y is shown as a function of u (a dimensionless variable):

$$u = \frac{x - a_3}{a_2}$$

(5.7.8)

As shown in 5.7.4 we assume that the data will be centered near the location of the peak (i.e., at $u = 0$). In other words, u_{min} will be approximately equal to $-u_{max}$. Results are shown in Figure 5.7.5 for Ω_k (k = 1, 2 and 3) versus u_{max}. The Ω_k's are defined as follows:

$$\Omega_1 = \frac{\sigma_{a1}(na_1)^{1/2}}{a_1} \qquad (5.7.9)$$

$$\Omega_k = \frac{\sigma_{ak}(na_1)^{1/2}}{a_2} \quad \text{for } k = 2 \text{ and } 3. \qquad (5.7.10)$$

As an example of the usage of these results, consider an experiment in which we want to measure a_3 (the peak location) to an accuracy of 0.01. The value is somewhere near $x = 20$ and our data ranges from 15 to 25. The approximate value of a_2 is 2 therefore the value of u_{max} is (25-20)/ $2 = 2.5$. Although it is difficult to read the value of Ω_3 from Figure 5.7.5 it appears to be about 1.2. (The result from an actual simulation for this experiment is 1.19). Assuming the experiment will be performed using a 256 channel analyzer, what value of a_1 is required to obtain the desired accuracy of 0.01? From Equation 5.7.10 :

$$a_1 = \frac{(\Omega_3 a_2)^2}{n\sigma_{a3}^2} = \frac{(1.19 * 2)^2}{256 * 0.01^2} = 221 \qquad (5.7.11)$$

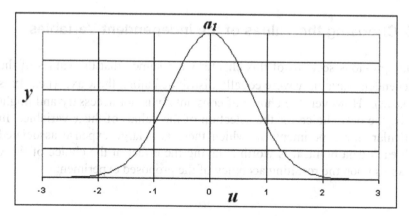

Figure 5.7.4   A Gaussian Peak with no Background

These results indicate that the experiment should be run long enough so that about 221 counts are recorded in the central channel. We can also ask how accurately the results for a_1 and a_2 should be in the same experiment. Since Ω_2 and Ω_3 are approximately the same, the value of σ_{a2} should also be about 0.01. The value of Ω_1 for this experiment is 2.07 so from Equation 5.7.9 we estimate that σ_{a1} should be about $2.07 * 221^{1/2}/256^{1/2} = 1.92$.

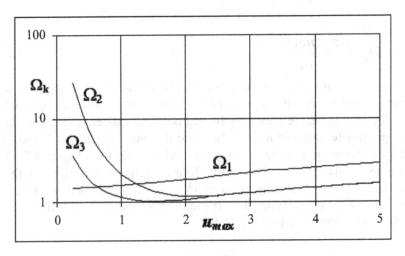

Figure 5.7.5 Ω_k versus u_{max}

5.8 Choosing the Values of the Independent Variables

In the previous sections of this chapter we assumed that the values of the independent variable x were equally distributed along the x axis (i.e., Δx is constant). However, the choice of constant Δx is not necessary and might not be the best choice for the selection of the values of the x variable. In particular, for experiments in which there is a large expense associated with each data point, it is worth studying the effect of the choice of the x variables upon the resulting accuracy of the proposed experiment.

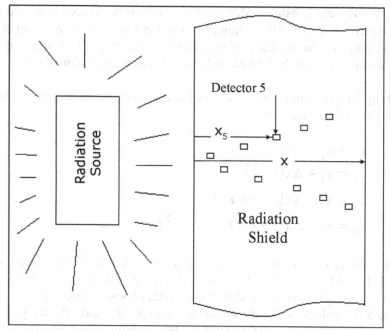

Figure 5.8.1   Experiment to Measure Effectiveness of a Radiation Shield

As an example, consider an experiment to measure the effectiveness of a radiation protection shield. The experiment will be run by placing 10 detectors within the shield. The radiation level will be measured at the 10 detector points simultaneously by recording radiation pulses at the detectors. The total number of counts at detector i located at point x_i is y_i with uncertainty $\sigma_{y_i} = y_i^{1/2}$. The experimental setup is shown in Figure 5.8.1. The applicable mathematical model is similar to Equation 5.7.3 but for this experiment the independent variable is x measured in units of length (cm) rather than t time units (e.g., sec):

$$y = f(x) = a_1 \cdot e^{-a_2 \cdot x} \tag{5.8.1}$$

We are interested in determining the values of both a_1 and a_2. The units of a_2 are in cm^{-1} and the intensity a_1 is in units of *counts*. Upon completion of the experiment the value of a_1 can be converted into a count rate by dividing the measured value by the duration of the experiment. Let us assume that the closest that we can get to the left side of the shield is 1 cm. If we were to choose a constant value of $\Delta x = 1$ then the detectors would be placed at $x = 1, 2, 3, .. 10$. If, however, we don't insist upon constant Δx, can we improve the accuracy of the proposed experiment? A preliminary measurement using only 2 points indicates that a_2 is approximately 0.5.

To study the problem of placement of the detectors, let us assume the following model starting from x_1:

$$x_2 = x_1 + \Delta x$$
$$x_3 = x_1 + \Delta x(1 + r)$$
$$x_4 = x_1 + \Delta x(1 + r + r^2)$$
$$x_n = x_1 + \Delta x(1 + r + r^2 + ...r^{n-2})$$

If we wish to have more points closer to the left side of the shield, we would choose a value of $r > 1$. Conversely, if we wish to have more points closer to the right side of the shield, we would choose a value of $r < 1$. In Figure 5.8.2 values of the dimensionless groups Ψ_1 and Ψ_2 are plotted for various combinations of r and Δx. Ψ_2 is defined in Equation 5.7.4. Ψ_1 is defined as follows:

$$\Psi_1 = \frac{\sigma_{a1}}{a_1}(a_1 n)^{1/2} \qquad (5.8.2)$$

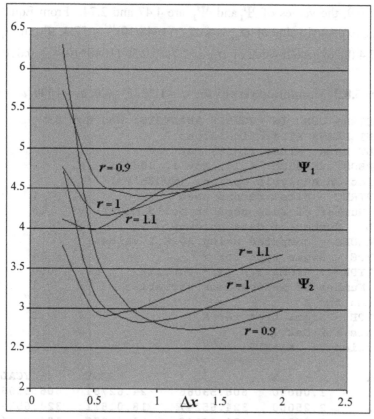

Figure 5.8.2 Ψ_1 and Ψ_2 versus Δx for 3 values of r

These results were obtained using simulations with the REGRESS program. The simulation output for $\Delta x = 1.25$ and $r = 0.9$ is shown in Figure 5.8.3. The curves in Figure 5.8.2 indicate that there is an optimum value of Δx that minimizes Ψ_1 and another optimum value that minimizes Ψ_2 for each value of r. However, comparing the results in the table for $r = 0.9$ and 1.1 to $r = 1$ (i.e., constant Δx), we note that the effect of r upon both Ψ_1 and Ψ_2 is not very dramatic.

As an example of the usage of these results, let us design the experiment so that we run it long enough so that 1000 counts are recorded at the point $x = 1$. From Figure 5.8.3 we see that if $a_1 = 1000$, then the number of counts at $x = 1$ is about 607 (regardless of the choice of Δx and r), so the value of a_1 should be increased by a factor of 1000 / 607 to 1648. If we use $\Delta x = 1.25$, and $r = 0.9$, the values of Ψ_1 and Ψ_2 are 4.47 and 2.74. From Equation 5.8.2 the expected value of σ_{a_1} is $4.47 * (1648/10)^{1/2} = 57.4$. From Equation 5.7.4 the expected value of σ_{a_2} is $2.74 * 0.5 / (1648*10)^{1/2} = 0.0107$.

Figure 5.8.3 Simulation results for $x = 1.25$, $r = 0.9$, $a_1 = 1000$, $a_2 = 0.5$

```
PARAMETERS USED IN REGRESS ANALYSIS: Wed Nov 03
  INPUT PARMS FILE: fig583.par
  INPUT DATA  FILE: fig583.dat
  REGRESS  VERSION: 4.10, Nov 1, 2004
 Prediction Analysis Option   (MODE='P')
  STARTREC - First record used            :      1
  N - Number of recs used to build model  :     10
  NCOL - Number of data columns           :      1
  Y VALUES - computed using A0 & X values
  SYTYPE - Sigma type for Y               :      4
     TYPE 4: SIGMA Y = CY * sqrt(Y)     CY: 1.000
  M - Number of independent variables     :      1
  Column for X                            :      1
  SXTYPE - Sigma type for X               :      0
 Analysis for Set 1
  Function Y:  A1*EXP(-A2*X)
```

POINT	X	Y	SIGY	YCALC
1	1.00000	606.53066	24.62784	606.53066
2	2.25000	324.65247	18.01811	324.65247
3	3.37500	184.98140	13.60079	184.98140
4	4.38750	111.49785	10.55925	111.49785
5	5.29875	70.69538	8.40805	70.69538
6	6.11888	46.91396	6.84938	46.91396
7	6.85699	32.43572	5.69524	32.43572
8	7.52129	23.26873	4.82377	23.26873
9	8.11916	17.25627	4.15407	17.25627
10	8.65724	13.18573	3.63122	13.18573

K	A0(K)	A(K)	PRED_SA(K)
1	1000.00	1000.00	44.71479
2	0.50000	0.50000	0.01372

POINT	X	YCALC	PRED_SIGY
1	1.00000	606.53066	20.99456
2	2.25000	324.65247	8.74574
3	3.37500	184.98140	5.24246
4	4.38750	111.49785	3.98922
5	5.29875	70.69538	3.19238
6	6.11888	46.91396	2.56505
7	6.85699	32.43572	2.06701
8	7.52129	23.26873	1.67806
9	8.11916	17.25627	1.37711
10	8.65724	13.18573	1.14454

Figure 5.8.3 (cont) Results for $x = 1.25$, $r = 0.9$, $a_1 = 1000$, $a_2 = 0.5$

5.9 Some Comments about Accuracy

In this chapter expected or predicted accuracy is discussed. Design of experiments for several classes of well known experiments is facilitated by providing equations or graphs for predicted uncertainty. For experiments not covered in this chapter, the methodology for obtaining similar prediction tools is developed. What should be understood is the accuracy of these uncertainty predictions.

In Section 5.2 it is explained that if the uncertainty models for the individual data points are reasonable, the expected value of S is $n-p$ and $S / n-p$ is one so therefore the predicted value of the standard deviation of a_k is:

$$\sigma_{a_k} = (C_{kk}^{-1})^{1/2} \tag{5.2.1}$$

and not the actual value as formulated in Equation 2.5.1:

$$\sigma_{a_k} = (\frac{S}{n-p} C_{kk}^{-1})^{1/2} \tag{2.5.1}$$

As explained in Section 3.2 the value of S is χ^2 (chi-squared) distributed with a mean value of $n-p$, so in actuality, we can predict a range of values for the σ_{a_k}'s. To appreciate the magnitude of this range, values of $S / n-p$ are listed for various values of $n-p$ and α in Table 5.9.1. The parameter α is the fraction of the distribution above the values listed in the columns. Clearly, as $n-p$ increases the range decreases and therefore the predicted values for σ_{a_k} are more accurate. However, this begins to appear as over-kill! When planning most experiments all we usually need are "ball-park" estimates of the values for σ_{a_k} that we can expect from the experimental results. The values in this table can be extended to larger values of $k = n\text{-}p$ by using the normal approximation to the χ^2 distribution. The standard deviation of the distribution is approaches $\sqrt{2k}$ for large values of k so the standard deviation of S / k is $\sqrt{2k} / k = \sqrt{2 / k}$. For example, for $k = 100$, the standard deviation is $(2/100)^{1/2} = 0.141$. For a standard normal distribution the 99% and 1% limits are at -2.326 and $+2.326$ so the limits for S / k are $1 - 0.141 * 2.326 = 0.672$ and $1 + 0.141 * 2.326 = 1.328$ as listed in the table).

$n\text{-}p$	$\alpha{=}0.99$	$\alpha{=}0.95$	$\alpha{=}0.05$	$\alpha{=}0.01$
4	0.074	0.178	2.372	3.319
6	0.145	0.273	2.099	2.802
8	0.205	0.342	1.938	2.511
10	0.256	0.394	1.831	2.321
15	0.349	0.484	1.666	2.039
20	0.413	0.543	1.571	1.878
25	0.461	0.584	1.506	1.773
30	0.498	0.615	1.459	1.696
50	0.535	0.671	1.329	1.465
100	0.672	0.724	1.276	1.328

Table 5.9.1   Values of $S / n\text{-}p$ for combinations of $n\text{-}p$ and α

As an example of the usage of the results in this table, consider the experiment analyzed in Section 5.8. In the experiment it was proposed to gather radiation data at 10 detector locations and the expected accuracies for the values of σ_{a_1} and σ_{a_2} were predicted to be 57.4 and 0.0107. Using Table 5.9.1, we can add ranges to these values. The value of $n\text{-}p$ is 10 $- 2 = 8$ so we can expect that if the experiment is repeated many times, 90% of the values of $S / n\text{-}p$ will be in the range 0.342 to 1.938. Substituting these values into Equation 2.5.1 the resulting 90% range for σ_{a_1} would be 19.6 to 111 and the resulting range for σ_{a_2} would be 0.00365 to 0.0207.

In other words, the ratio of the maximum and minimum values of these ranges is greater than a factor of 5 (i.e., 1.938 / 0.342 = 5.67). We see that for this experiment the predicted values of σ_{a_1} and σ_{a_2} are really just ball-park estimates!

Chapter 6 SOFTWARE

6.1 Introduction

One of the earliest applications of digital computers was least squares analysis of experimental data. The Manhattan Project during World War II included a large emphasis on experiments to determine basic properties such as half lives of radioactive isotopes, radiation shielding parameters, biological effects of radiation and many other properties of vital interest. The fundamentals of nonlinear least squares analysis was known then and was summarized in a book by W. E. Deming in 1943 [DE43]. An unclassified Los Alamos publication in 1960 by R. Moore and R. Zeigler described the software used at Los Alamos for solving nonlinear least squares problems [MO60]. Besides describing their general purpose software, they discussed some of the problems encountered in converging to a solution for some mathematical models.

Most readers of this book are either users of available NLR (nonlinear regression) software or are interested in evaluating and/or obtaining NLR software. Some readers, however, will be interested in writing their own software to solve a specific problem. Chapter 2 includes sufficient details to allow a user to rapidly get a system up and running. For all readers it should be useful to survey features that one would expect to see in a general purpose NLR program. It should be emphasized that there is a difference between a general purpose NLR program and a program written to quickly solve a specific problem. Indeed, using a language like MATLAB, some of my students in a graduate course that I have been teaching for a number of years (Design and Analysis of Experiments) have produced NLR code to solve specific homework problems.

Statistical software is available through the internet from a massive variety of sources. A Google search for "statistical software" turned up 9.5 million hits! Some of the software is free and other software programs are available for a price that can vary over a wide range. Some of the software includes nonlinear regression applications. Refining the search by adding "nonlinear regression" turned up over 600,000 hits. Many of these hits describe nonlinear regression modules that are part of larger statistical packages. Further refining the search to S-plus, the number of hits was over 26,000. Nonlinear regression software in S-plus is described by Venables and Ripley [VE02]. Huet et. al. describe a program called NLS2 that runs under the R statistical software environment as well as S-plus [HU03]. Refining the search to SPSS, the number of hits was over 30,000. The SPSS Advanced Statistics Manual includes details for nonlinear regression analyses within SPSS [ZE98]. Refining the search to SAS, the number of hits was about 51,000. The NLIN procedure in the SAS system is a general purpose nonlinear regression program and is described in a paper by Oliver Schabenberger [SC98]. Refining the search to MATLAB, over 41,000 hits were noted. MATLAB *m* files for performing nonlinear regression analyses are included in [CO99]. The MATLAB Statistical Toolbox includes a function called *nlinfit* for performing nonlinear regression [www.mathworks.com/products/statistics].

In Section 6.2 features that are common to general purpose NLR programs are described and features that are desirable but not available in all the programs are also described. In Section 6.3 the NIST Statistical Reference Datasets are discussed. These well-known datasets are used to evaluate NLR programs and search algorithms. In Section 6.4 the subject of convergence is discussed. For most users, performance of NLR programs is primarily based upon a single issue: does the program achieve convergence for his or her problems of interest? In Section 6.5 a problem associated with linear regression is discussed. Multi-dimensional modeling is discussed in Section 6.6 and software performance is discussed in Section 6.7.

6.2 General Purpose Nonlinear Regression Programs

There are a number of general purpose nonlinear regression programs that can easily be obtained and allow the user to run most problems that he or she might encounter. Some of the programs are offered as freeware and some of the programs must be purchased. Some programs are offered on a free trial basis but must then be purchased if the user is satisfied and wishes to use the program after termination of the trial period. This section includes a survey of features that one encounters while reviewing nonlinear regression software. The purpose of this chapter is to provide the reader with the necessary background required to make a reasoned choice when deciding upon which program to use for his or her specific applications.

When one works within the framework of a general purpose statistical software environment (e.g., SAS, SPSS, S-plus, MATLAB Statistical Toolbox), a reasonable choice for nonlinear regression is a program that is compatible with the environment. Data created by one module of the system can then be used directly by the nonlinear regression module. Alternatively one can use a general purpose nonlinear regression program that runs independently (i.e., not within a specific statistical software environment). One problem with this alternative is data compatibility but this need not be a major obstacle. Most statistical software environments are Excel compatible, so if the nonlinear regression program is also Excel compatible, then data can be easily moved from the statistical software environment through Excel to the nonlinear regression program. In addition, ASCII text files can be used by almost all general purpose programs and statistical environments.

To qualify as a general purpose nonlinear regression program I feel that as a minimum, the following features should be included:

1) Mathematical models should be entered as input parameters.
2) The program should accept nonlinear models with respect to the unknown parameters (and not just nonlinear with respect to the independent variables).
3) There should be no need for the user to supply derivatives (neither analytical nor numerical) of the mathematical model.
4) The program should be able to accommodate mathematical models that are multi-dimensional in both the dependent and independent variables.
5) The user should be able to weight the data according to any weighting scheme of his or her choosing.
6) The program should include a sophisticated convergence algorithm. The National Institute of Standards nonlinear regression datasets (described in Section 6.3) provide a rich variety of problems that can be used to test the quality of a program's ability to achieve convergence.

In addition, there are a number of desirable features that one would like to see in a general nonlinear regression program:

1) Allow the user to name the dependent and independent variables and the unknown parameters.
2) Allow the user to define symbolic constants.
3) Allow specification of Bayesian estimators for the unknown parameters.
4) Include a simulation feature useful for designing experiments. (See Chapter 5 for a discussion and examples related to this feature.)
5) Allow input of Excel text files.
6) Include a feature to generate an interpolation table that lists values of the dependent variable and their standard deviations for a specified set of the independent variable or variables.
7) Allow program usage from within a general purpose statistical or programming environment.
8) Include a feature for generation of graphical output.

Treating mathematical models as input parameters is probably the most important feature of a general purpose NLR program. If the user is forced to program a function for every problem encountered, then the NLR program is not really "general purpose". If the user is working in an interactive mode and notes that a particular function does not yield results of sufficient accuracy, he or she should be able to enter a new function without having to exit the NLR program to reprogram the function.

The need for symbolic constants is a feature that can be most useful for problems in which convergence is difficult. This subject is discussed in Section 6.4.

There are several debatable features that are really a matter of user preference. Should the program use a parameter file for specifying the parameters of a particular analysis or should the program work through a GUI interface? Today, most computer programs (not just nonlinear regression programs) are interactive and allow the user to specify what he or she wishes to do thru a menu driven series of questions. For nonlinear regression, the number of parameters can be considerable so if the program is accessed through a GUI interface, there should be some method for shortcutting the process when the change from a previous analysis is minor. This particular problem is avoided if parameter files are used. All one has to do is edit the parameter file and make changes or perhaps copy the file under a new name and then change the new file.

The need for graphic output is a very reasonable user requirement, but should it be an integral part of an NLR general purpose program? As long as one can easily port the data to another program that supports graphics, then this should be a reasonable compromise. For example, if the NLR program outputs text data, this output can then be inputted to a program like Excel to obtain the necessary graphical output.

6.3 The NIST Statistical Reference Datasets

The U.S. National Institute of Standards and Technology (NIST) initiated a project to develop a standard group of statistical reference datasets (StRD's). In their words the object of the project was "to improve the accuracy of statistical software by providing reference datasets with certified computational results that enable the objective evaluation of statistical software." One of the specific areas covered was datasets for nonlinear regression. The NIST StRD project home page can be accessed at:

http://www.itl.nist.gov/div898/strd/index.html

To examine the datasets, go into *Dataset Archives* and then *Nonlinear gression.* A summary of the NIST nonlinear regression datasets is cluded in Table 6.3.1:

Name	Difficulty	Parms	Num pts	Function
Misrala	Lower	2	14	b1*(1-exp[-b2*x])
Chwirut1	Lower	3	214	exp[-b1*x]/(b2+b3*x)
Chwirut2	Lower	3	54	exp(-b1*x)/(b2+b3*x)
Lanczos3	Lower	6	24	b1*exp(-b2*x) + b3*exp(-b4*x) + b5*exp(-b6*x)
Gauss1	Lower	8	250	b1*exp(-b2*x) + b3*exp(-(x-b4)**2 / b5**2) + b6*exp(-(x-b7)**2 / b8**2)

Name	Difficulty	Parms	Num pts	Function
Gauss2	Lower	8	250	Same as Gauss1
DanWood	Lower	2	6	b1*x**b2
Misralb	Lower	2	14	b1 * (1-(1+b2*x/2)**(-2))
Kirby2	Average	5	151	(b1 + b2*x + b3*x**2) / (1 + b4*x + b5*x**2)
Hahn1	Average	7	236	(b1+b2*x+b3*x**2+b4*x**3) / (1+b5*x+b6*x**2+b7*x**3)
Nelson	Average	3	128	b1 - b2*x1 * exp[-b3*x2]
MGH17	Average	5	33	b1 + b2*exp[-x*b4] + b3*exp[-x*b5]
Lanczos1	Average	6	24	b1*exp(-b2*x) + b3*exp(-b4*x) + b5*exp(-b6*x)
Lanczos2	Average	6	24	Same as Lanczos1
Gauss3	Average	8	250	Same as Gauss1
Misralc	Average	2	14	b1 * (1-(1+2*b2*x)**(-.5))
Misrald	Average	2	14	b1*b2*x*((1+b2*x)**(-1))
Roszman1	Average	4	25	b1 - b2*x - arctan[b3/(x-b4)]/pi
ENSO	Average	9	168	b1 + b2*cos(2*pi*x/12) + b3*sin(2*pi*x/12) + b5*cos(2*pi*x/b4) + b6*sin(2*pi*x/b4) + b8*cos(2*pi*x/b7) + b9*sin(2*pi*x/b7)
MGH09	Higher	4	11	b1*(x**2+x*b2) / (x**2+x*b3+b4)
MGH10	Higher	3	16	b1 * exp[b2/(x+b3)]
Thurber	Higher	7	37	(b1 + b2*x + b3*x**2 + b4*x**3) /

```
NIST/ITL StRD
Dataset Name:   BoxBOD              (BoxBOD.dat)
Description:   These data are described in detail in
Box, Hunter and Hunter (1978).  The response variable
is biochemical oxygen demand (BOD) in mg/l, and the
predictor variable is incubation time in days.

Reference: Box,G.P., W.G.Hunter, and J.S.Hunter(1978).
           Statistics for Experimenters.
           New York, NY: Wiley, pp. 483-487.


Data:      1 Response  (y = biochemical oxygen demand)
           1 Predictor (x = incubation time)
           6 Observations
           Higher Level of Difficulty
           Observed Data

Model:     Exponential Class
           2 Parameters (b1 and b2)

           y = b1*(1-exp[-b2*x])   +   e
Start 1 Start 2    Parameter       Standard Deviation
b1=1      100    2.1380940889E+02  1.2354515176E+01
b2=1      0.75   5.4723748542E-01  1.0455993237E-01

Residual Sum of Squares:        1.1680088766E+03
Residual Standard Deviation:    1.7088072423E+01
Degrees of Freedom:             4
Number of Observations:         6

Data:   y              x
        109            1
        149            2
        149            3
        191            5
        213            7
        224            10
```

Figure 6.3.1   Data and Results for NIST Dataset BoxBOD

Name	Difficulty	Parms	Num pts	Function
BoxBOD	Higher	2	6	$(1 + b5*x + b6*x**2 + b7*x**3)$ $b1*(1-exp[-b2*x])$
Ratkwosky3	Higher	3	9	$b1 / (1+exp[b2-b3*x])$
Ratkowsky4	Higher	4	15	$b1 / ((1+exp[b2-b3*x])**(1/b4))$
Eckerle4	Higher	3	35	$(b1/b2) *$ $exp[-0.5*((x-b3)/b2)**2]$
Bennett5	Higher	3	154	$b1 * (b2+x)**(-1/b3)$

Table 6.3.1   Datasets from the NIST Nonlinear Regression Library

There are 27 different data sets included in the library including the actual data files (in text format) and results from least squares analyses of the data. The datasets are classified by Level of Difficulty (lower, average and higher), number of parameters (varying from 2 to 9), and number of data points (varying from 6 to 250). The datasets including the mathematical models are listed in Table 6.3.1. Each data set includes two different starting points: one near the solution and one further away from the solution. Also included are the least squares values and their estimated standard deviations. Results also include the sum of the squares of the residuals (i.e., *S*) and the Residual Standard deviation (i.e., *sqrt(S / (n-p))*). The datasets include a number of challenging problems that test the ability of a program to converge to a solution. However, the choice of datasets is limited to mathematical models that include a single independent variable *x*. Another limitation is that only unit weighting is used for the all problems. Details for one of the datasets (BoxBOD) are shown in Figure 6.3.1.

The BoxBOD problem has only two unknown parameters (i.e., b1 and b2) and only six data points and yet it is listed as *higher* in level of difficulty because of the difficulty of converging to a solution from the Start 1 initial values.

One of the most well-known general purpose nonlinear regression programs is **NLREG** (www.nlreg.com). They describe their results using the NIST datasets as follows: "The NIST reference dataset suite contains 27 datasets for validating nonlinear least squares regression analysis software. **NLREG** has been used to analyze all of these datasets with the following results: **NLREG** was able to successfully solve 24 of the datasets, producing results that agree with the validated results within 5 or 6 significant digits. Three of the datasets (Gauss1, Gauss2 and Gauss3) did not converge, and NLREG stopped with the message: *Singular convergence. Mutually dependent parameters?* The primary suggested starting values were used for all datasets except for MGH17, Lanczos2 and BoxBOD which did not converge with the primary suggested starting values but did converge with the secondary suggested starting values."

The differences between the three Gauss datasets are in the data. A plot of the Gauss1 data is shown in Figure 6.3.2. All three models include two Gaussian peaks with an exponentially decaying background. The peaks in the Gauss3 dataset are much closer together than the peaks in the other two datasets and that is why its level of difficulty is considered higher. However, NIST lists all three datasets as either lower or average level of difficulty. I ran Gauss1, Gauss2 and Gauss3 using **REGRESS** and had no problem converging from the primary suggested starting values. My guess is that somehow an error was introduced in the **NLREG** tests for these three datasets because it isn't logical that **NLREG** would fail for these and yet pass all the tests for the higher level of difficulty.

Another available NLR general purpose program is **LabFit** which can be located at http://www.angelfire.com/rnb/labfit/index.htm. Results for all the NIST datasets are included on the website. In their words they "achieved convergence for all the primary starting values for all the datasets and the results are statistically identical to the certified values".

Results for the NLR program **Stata** 8.1 can be seen at http://www.stata.com/support/cert/nist/. They achieved convergence for all of the datasets but for one dataset of average difficulty (MGH17) and 4 of higher difficulty (MGH09, MGH10, Eckerle4 and Ratkowsky4) they only achieved convergence from the nearer starting points. **Stata** is a "complete, integrated statistical package that provides everything needed for data analysis, data management, and graphics". The NLR module is only one feature of the **Stata** package.

Results for the program **Datafix** (a product of Oakdale Engineering) are available at http://www.curvefitting.com/datasets.htm. They achieved convergence for all datasets "without using analytical derivatives" but do not specify if this was from the primary or secondary starting points.

An Excel based NLR program is included as part of the **XLSTAT** package. Details can be obtained at http://www.xlstat.com/indexus.html. This package runs within Excel and they include the Ratkowsky4 example in their demonstration. Their solution requires programming of the derivatives of the modeling function and therefore cannot be considered as a general purpose NLR program. However, they have programmed complete solutions including derivatives for a limited number of functions.

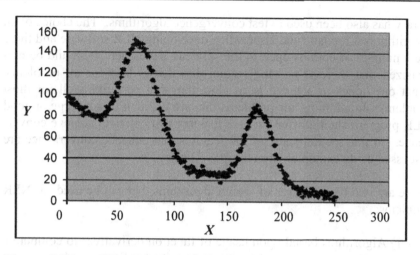

Figure 6.3.2   Gauss1 data from NIST Nonlinear Regression Library

An NLR program is included in the **TSP** econometrics package. The results for the NIST nonlinear reference datasets can be seen on the **TSP** International website:

http://www.tspintl.com/products/tsp/benchmarks/nlstab.txt

They achieved convergence on all the datasets except Lanczos1. No mention is made regarding the starting points for the various tests.

The **LIMDEP** program (a product of Econometric Software) is another general purpose statistical econometric package. Details regarding the **LIMDEP** program can be obtained at:

http://www.limdep.com/programfeatures_accuracy.shtml

The **LIMDEP** NLR module was tested using the NISP datasets as well as other benchmark datasets described by McCullough [MC99]. In their own words: "**LIMDEP** was able to solve nearly all the benchmark problems using only the program default settings, and all of the rest with only minor additional effort." This statement makes a lot of sense. Most general purpose NLR programs have default settings but for difficult problems, some minor adjustments in the parameters can lead to convergence. This subject is considered in Section 6.4.

6.4 Nonlinear Regression Convergence Problems

In Section 6.3 the NIST dataset library of NLR problems is discussed. The library has been used extensively to test NLR computer programs. The library has also been used to test convergence algorithms. The choice of an algorithm is a fundamental issue when developing NLR software and there are a number of options open to the software developer. It should be emphasized that there is no single algorithm that is best for all problems. What one hopes to achieve is an algorithm that performs well for most problems. In addition, for problems that are difficult to converge, a good NLR program should offer the user features that can help achieve convergence. In this section some of the features that enhance convergence are discussed using examples from the NIST library.

There are two basic classes of search algorithms that can be used for NLR problems:

1) Algorithms based upon usage of function derivatives to compute a vector of changes in the manner described in Section 2.4.
2) Stochastic algorithms that intelligently search thru a defined unknown parameter space.

The straight forward Gauss-Newton (GN) algorithm (Equations 2.4.16 and 2.4.17) is the starting point for most algorithms of the first type. This simple algorithm leads to convergence for many NLR problems but is not sufficient for more difficult problems like some of those encountered in the NIST datasets. To improve the probability of achieving convergence, Equation 2.4.16 can be replaced by:

$$a_k = a0_k + caf * A_k \qquad k = 1 \text{ to } p \qquad (6.4.1)$$

where *caf* is called the convergence acceleration factor. As a default *caf* is one, but for difficult problems, using a value of *caf* < 1 can sometimes lead to convergence. A more sophisticated approach is to calculate the value of S computed using the new values of a_k and compare this value with the value of S computed using the old values. As long as the value of S decreases, continue along this line (i.e., increase *caf*). However, if the reverse is true (i.e., $S_{new} > S_{old}$) the value of *caf* is decreased (even to a negative number). It should be emphasized that *caf* is an input parameter and all changes of *caf* should be done algorithmically within the program for a given iteration. For the next iteration the value of *caf* is restarted at the input value. Sometimes it turns out that along the direction suggested by the A vector, S rises in both directions (i.e., *caf* > 0 and *caf* < 0). When this happens the algorithm can be modified to alter the direction. The Marquardt algorithm (sometimes called the Levenberg-Marquardt algorithm) is very popular and is used to modify the basic Gauss Newton algorithm [LE44, MA63, GA92]. Equations 2.4.16 or 6.4.1 are still used but the A vector is computed using a modified procedure. Instead of computing the A vector using Equation 2.4.9, the following equation is used:

$$A = (C + \lambda D)^{-1} V \qquad (6.4.2)$$

The matrix D is just the diagonal of the C matrix (with all off diagonal terms set to zero) and λ is a scalar parameter. By trying several different values of λ a new direction can often be found which leads to a better reduction of S then achieved using Equation 2.4.16 or 6.4.1.

Tvrdik and Krivy survey several standard algorithms using the higher difficulty problems from the NIST datasets [TV04]. This paper can also be accessed online at http://albert.osu.cz/tvrdik/down/files/comp04.pdf. The algorithms used are those included in several NLR standard packages: NCSS 2001 which uses a Levenberg-Marquardt (LM) algorithm, S-PLUS 4.5 which uses a GN algorithm, SPSS 10.0 which uses a modified LM algorithm and SYSTAT 8.0 which includes both a modified GN algorithm and an algorithm based upon the simplex method. Their results are shown in Table 6.4.1.

	NCSS		SYST GN		SYST Sim		S-Plus		SPSS	
Start:	1	2	1	2	1	2	1	2	1	2
Bennett5	2	1	OK	OK	F	F	OK	OK	OK	OK
BoxBOD	F	F	OK	OK	F	F	OK	OK	F	OK
Eckerle4	F	3	OK	OK	F	F	F	OK	OK	OK
MGH09	F	F	F	OK	OK	OK	F	2	OK	OK
MGH10	F	F	OK	OK	F	OK	F	OK	F	F
Ratkowsky3	OK	OK	OK	OK	F	F	F	OK	OK	OK
Ratkowsky4	F	3	OK	OK	F	F	F	OK	OK	OK
Thurber	F	F	OK	OK	OK	OK	F	F	F	F

Table 6.4.1 Comparison of algorithms for NIST datasets.

For each dataset, the programs were started from the far (1) and near (2) points as listed in the NIST reference datasets. An entry of F means that the program failed to converge and OK means that it did converge and S was accurate to at least 4 significant digits. A numerical entry means that it converged to 1, 2 or 3 significant digits. Clearly the SYSTAT program using the modified GN algorithm outperformed the other program but this does not mean that a GN algorithm is necessarily best. It does, however, prove that by cleverly modifying the basic algorithm one can achieve better results.

One of the easiest features that can be employed in an NLR program is to limit the search for some or all of the unknown parameters. For example, consider the BoxBOD dataset from the NIST library. Details are shown in Figure 6.3.1. Results obtained using the REGRESS program with only the default parameters are shown in Figure 6.4.1. An examination of the results shows that the value of **B2** becomes a huge negative number. Looking at the data in Figure 6.3.1 and the function used to specify **Y** we see that **Y** increases with **X** so **B2** must be a positive number. Setting a value of **B2MIN** = 0.001 and rerunning the program, the results in Figure 6.4.2

are obtained after 80 iterations. The ability to specify minimum and maximum values for the unknown parameters is an essential feature in a general purpose NLR program.

```
PARAMETERS USED IN REGRESS ANALYSIS: Thu Dec 02, 2004
    INPUT PARMS FILE: boxbod.par
    INPUT DATA  FILE: boxbod.par
    REGRESS  VERSION: 4.10, Nov 15, 2004
        STARTREC - First record used              :     1
        N - Number of recs used to build model    :     6
        NO_DATA - Code for dependent variable    -999.0
        NCOL - Number of data columns             :     2
        NY   - Number of dependent variables      :     1
        YCOL1 - Column for dep var Y              :     1
        SYTYPE1 - Sigma type for Y                :     1
          TYPE 1:  SIGMA Y = 1
        M - Number of independent variables       :     1
        Column for X1                             :     2
        SXTYPE1 - Sigma type for X1               :     0
          TYPE 0:  SIGMA X1 = 0
    Analysis for Set 1
        Function Y:   B1*(1-EXP[-B2*X])
        EPS - Convergence criterion        :  0.00100
        CAF - Convergence acceleration factor :    1.000

    ITERATION          B1          B2    S/(N.D.F.)
            0       1.00000     1.00000    46595.60
            1      89.08912   114.70610    12878.94
            2     185.20000    <-10^49      >10^09
    Singular matrix condition
```

Figure 6.4.1   Results for BoxBOD using Default Settings

There are some problems in which the values of the unknown parameters vary slowly but convergence is very difficult to achieve. For such problems setting upper and lower bounds on the parameters accomplishes nothing. The Bennett5 problem from the NIST datasets is an example of such a problem. Using the far starting points for the 3 unknowns, REGRESS required over 536,000 iterations to achieve convergence! Using the near starting points the results were not much better: over 390,000 iterations were required. REGRESS uses a modified GN algorithm but if the progress for an iteration is not sufficient it then uses an LM algorithm. A better approach for problems of this type is to use a stochastic algorithm. Stochastic algorithms avoid the need for function derivatives. A search space is defined by setting minimum and maximum values for all the unknown parameters. A random number generator is used to set a starting point within the space and then a heuristic is used to find the next point. In the same paper as mentioned above [TV04], Tvrdik and Krivy describe 5 different stochastic algorithms and then compare them using the same datasets as listed in Table 6.4.1. Their results show large performance differences from problem to problem and algorithm to algorithm. Four of the five managed to achieve solutions for the Bennett5 problem.

```
PARAMETERS USED IN REGRESS ANALYSIS: Thu Dec 02, 2004
      ITERATION           B1            B2    S/(N.D.F.)
             0       1.00000       1.00000      46595.60
             1      89.08912     114.70610      12878.94
             2     185.20000       0.00100      46567.68
             3       9985.49       0.05420    7946907.22
             4      -1977.29       0.07206     917128.67
             5    -907.83514       0.00172      51485.53
             6       7854.00       0.00100      28275.45
             7      15098.02       0.00193       6215.57
             8      14635.85       0.00203       6194.60
        - - - - - - - - - - - - - - - - - - - -
        - - - - - - - - - - - - - - - - - - - -
            79     213.87781       0.54643     292.00568
            80     213.82425       0.54706     292.00238
POINT           X1            Y          SIGY         YCALC
    1      1.00000    109.00000      1.00000      90.10764
    2      2.00000    149.00000      1.00000     142.24097
    3      3.00000    149.00000      1.00000     172.40360
    4      5.00000    191.00000      1.00000     199.95134
    5      7.00000    213.00000      1.00000     209.17267
    6     10.00000    224.00000      1.00000     212.91397

PARAM INIT_VALUE  MINIMUM   MAXIMUM    VALUE       SIGMA
   B1    1.00000 Not Spec  Not Spec 213.81258    12.35691
   B2    1.00000  0.00100  Not Spec   0.54720     0.10452
Variance Reduction:          88.05
S/(N - P)           :     292.00223
RMS (Y - Ycalc)     :      13.95235
```

Figure 6.4.2   Results for BoxBOD using B2MIN = 0.001

Another option for problems that are difficult to converge is to use symbolic constants. For example, the parameter file for the REGRESS runs for the Bennett5 problem included the following function specification:

```
unknown b1, b2, b3;
y ='b1 * (b2+x)^(-1/b3)'
```

Knowing the solution in advance, and noticing that the values of the un-knowns were progressing in the correct direction, I just let REGRESS run until convergence was achieved. However, if the amount of data had been much greater than the 154 data records associated with this dataset, the time required to reach convergence would have been very large indeed. An alternative to this approach is to use symbolic constants. For example, one could hold **b1** constant and do a two parameter fit using the following function specification:

```
constant b1;
unknown b2, b3;
y ='b1 * (b2+x)^(-1/b3)'
```

Once least square values of **b2** and **b3** are located for the inputted value of **b1** the value can be changed and a new combination can be located. Com-paring the values of S obtained for the different values of **b1**, one can home in on a region likely to contain the best value of **b1**. Once this re-gion has been identified, one could then return to the original function specification to make the final 3 parameter search. The number of itera-tions using this procedure is much less than starting the process searching for all 3 parameters but requires a lot of user intervention and judgment.

For very difficult problems a combination approach is sometimes used. The process is started by doing a very course grid search through the entire space just computing S at all points in the grid. The best region to start the search is around the point for which S is a minimum. All the unknowns are then bounded within this region and a detailed search is then initiated. If convergence is still a problem, then the use of symbolic constants and/or a stochastic algorithm can be used to further reduce the size of the search space.

6.5 Linear Regression: a Lurking Pitfall

A general purpose NLR (nonlinear regression) program can easily handle linear regression problems. Software developed for nonlinear problems can be used with no change to solve linear problems. However, there is a hidden danger in using linear models that often plagues new users of curve-fitting software. When data is available and there is no physically meaningful mathematical model to explain the variation of a dependent variable y as a function of x, the most tempting approach to the problem is to use a simple polynomial:

$$y = a_1 + a_2 x + a_3 x^2 + \ldots + a_p x^{p-1} \tag{6.5.1}$$

If one is only looking for an adequate function to predict y for any value of x then why not just start with a straight line (i.e., $p = 2$) and increase p until the average root-mean-square (RMS) error is acceptable? This approach, although theoretically very appealing, can lead to very difficult numerical problems that arise due to the fact that computers work to a finite number of significant digits of accuracy.

To explain the problem, consider data in which the values of x are equally spaced from 0 to 1 and unit weighting is used. The derivative of Equation 6.5.1 with respect to a_k is simply x^{k-1} so from Equations 2.4.14 and 2.4.15 the terms of the C matrix and the V vector are:

$$C_{jk} = \sum_{i=1}^{i=n} \frac{\partial f}{\partial a_j} \frac{\partial f}{\partial a_k} = \sum_{i=1}^{i=n} x^{j-1} x^{k-1} = \sum_{i=1}^{i=n} x^{j+k-2} \tag{6.5.2}$$

$$V_k = \sum_{i=1}^{i=n} Y_i \frac{\partial f}{\partial a_k} = \sum_{i=1}^{i=n} Y_i x^{k-1} \tag{6.5.3}$$

Once we have computed the terms of the C matrix and the V vector we use Equation 2.4.9 to solve for the vector A:

$$A = C^{-1}V \tag{2.4.9}$$

This vector includes all p values of the a_k's. We can estimate the value of C_{jk} by using the following approximation:

$$C_{jk} = \sum_{i=1}^{i=n} x^{j+k-2} = n\left(x^{j+k-2}\right)_{avg} \approx n \int_0^1 x^{j+k-2} dx = \frac{n}{j+k-1} \tag{6.5.4}$$

For example, for $p = 4$ the C matrix is approximately:

$$C = n\begin{bmatrix} 1 & 1/2 & 1/3 & 1/4 \\ 1/2 & 1/3 & 1/4 & 1/5 \\ 1/3 & 1/4 & 1/5 & 1/6 \\ 1/4 & 1/5 & 1/6 & 1/7 \end{bmatrix} \qquad (6.5.5)$$

For those readers familiar with linear algebra, they will recognize this matrix as the well known Hilbert matrix and it has the following property:

$$cond(C) \approx e^{3.5p} = 10^{3.5p/ln(10)} \equiv 10^{1.5p} \qquad (6.5.6)$$

In other words, as the number of unknowns (i.e., p) increases, the condition of the matrix (the ratio of the largest to smallest eigenvalues of the matrix) increases exponentially. Using **cond(C)** we can estimate the errors in the a_k's due to errors from the terms of the V vector:

$$\frac{\|\delta A\|}{\|A\|} \le cond(C)\frac{\|\delta V\|}{\|V\|} \qquad (6.5.7)$$

This equation means that the fractional errors in the terms of the A vector are no more than **cond(C)** times the fractional errors in the V vector. For example, let us assume that the values of Y are accurate to 5 decimal digits so that the fractional errors in the terms of the V vector are of the order of 10^{-5}. If **cond(C)** is about 100, then the fractional errors in the terms of the A vector are at worst of the order of 10^{-3}. This loss of accuracy comes about due to the process of inverting the C matrix. In other words, if **cond(C)** is about 100 we can expect a loss of about 2 digits of accuracy in solving Equation 2.4.6 (i.e., $CA = V$). A set of linear equations like Equation 2.4.6 is said to be "ill-conditioned" when the value of the condition become a large number.

Examining Equations 6.5.6 and 6.5.7, the pitfall in using Equation 6.5.1 for curve fitting can be seen. As p increases, C becomes increasingly ill-conditioned. The log_{10} of **cond(C)** is the maximum number of decimal digits that might be lost in solving Equation 2.4.6. So if $p = 5$, 6 or 7 then the condition is $10^{7.5}$, 10^9 or $10^{10.5}$ and the number of digits of accuracy that might be lost are 7.5, 9 or 10.5! We see that even though Equation 6.5.1 is a very tempting solution for obtaining a simple equation relating y to x, it is increasingly numerically problematical as p increases.

The NIST datasets include linear as well as nonlinear problems. The most difficult problem is the 'Filippelli problem'. This dataset has 82 point and the proposed model is Equation 6.5.1 with $p = 11$. The LIMDEP website includes their solution to this problem and they describe the problem as follows:

> "LIMDEP's linear regression computations are extremely accurate. The 'Filippelli problem' in the NIST benchmark problems is the most difficult of the set. Most programs are not able to do the computation at all. The assessment of another widely used package was as follows: Filippelli test: XXXXX found the variables so collinear that it dropped two of them – that is, it set two coefficients and standard errors to zero. The resulting estimates still fit the data well. Most other statistical software packages have done the same thing and most authors have interpreted this result as acceptable for this test. We don't find this acceptable. First, the problem is solvable. See LIMDEP's solution below using only the program defaults - just the basic regression instruction. Second, LIMDEP would not, on its own, drop variables from a regression and leave behind some arbitrarily chosen set that provides a 'good fit.' If the regression can't be computed within the (very high) tolerance of the program, we just tell you so. For this problem, LIMDEP does issue a warning, however. What you do next is up to you, not the program."

It should be emphasized that the Filippelli problem is a problem that was proposed to test software and not a real problem in which Mr. Filippelli was actually trying to get usable numbers. If one proceeds using Equation 6.5.1 directly, consider the loss of accuracy using a 10th order polynomial (i.e., $p = 11$) to fit the data. The number of digits of accuracy lost is at a maximum 16.5! Even if the values of Y are true values with no uncertainty, just inputting them into double precision numbers in the computer limits their accuracy to about 15 digits. So a loss of 16.5 digits makes the results completely meaningless. The C matrix is so ill-conditioned that it is no wonder that most packages fail when trying to solve the Filippelli problem. I tried running this problem using REGRESS and could not progress beyond $p = 9$.

So how did LIMDEP succeed while others have failed? I don't know the algorithm used by LIMDEP to solve problems based upon Equation 6.5.1, but if I was interested in creating software to solve such problems I would use orthogonal polynomials [RA78, WO71]. The idea originally proposed by G. Forsythe [FO57] is to replace Equation 6.5.1 with the following:

$$y = \sum_{k=0}^{k=p} a_k u_k(x) \qquad (6.5.8)$$

The $u_k(x)$ terms are a set of p polynomials all orthogonal to one another. Orthogonality for a particular set of data and a particular weighting scheme implies the following:

$$\sum_{i=1}^{i=n} w_i u_j(x_i) u_k(x_i) = 0 \qquad \text{for } j \neq k. \qquad (6.5.9)$$

Equation 2.4.5 is applicable to all linear models and is therefore applicable to Equation 6.5.8. Substituting u for g in Equation 2.4.5 we get $p+1$ equations of the following form (where the index k is from 0 to p):

$$a_0 \sum w_i u_0 u_k + a_1 \sum w_i u_1 u_k + ... + a_p \sum w_i u_p u_k = \sum w_i Y_i u_k \qquad (6.5.10)$$

Applying Equation 6.5.9 to 6.5.10 we end up with $p+1$ equations for a_k that can be solved directly:

$$a_k \sum w_i u_k u_k = \sum w_i Y_i u_k \quad k = 0 \text{ to } p \qquad (6.5.11)$$

$$a_k = \frac{\sum w_i Y_i u_k}{\sum w_i u_k u_k} \quad k = 0 \text{ to } p \qquad (6.5.12)$$

If a set of polynomials can be constructed with this property (i.e., Equation 6.5.9), then we can compute the terms of the A vector without inverting the C matrix. Or looking at it another way, the diagonal terms of the C^{-1} matrix are the inverses of the diagonal terms of the C matrix and all the off-diagonal terms are zero. Forsythe suggests the following scheme for computing polynomials satisfying Equation 6.5.9:

$$u_0(x) = 1 \qquad\qquad (6.5.13a)$$

$$u_1(x) = (x - \alpha_1)u_0(x) \qquad\qquad (6.5.13b)$$

$$u_2(x) = (x - \alpha_2)u_1(x) - \beta_1 u_0(x) \qquad\qquad (6.5.13c)$$

$$u_p(x) = (x - \alpha_p)u_{p-1}(x) - \beta_{p-1}u_{p-2}(x) \qquad\qquad (6.5.13d)$$

The α's and β's are computed as follows:

$$\alpha_k = \frac{\displaystyle\sum_{i=0}^{n} x_i w_i (u_{k-1}(x_i))^2}{\displaystyle\sum_{i=0}^{n} w_i (u_{k-1}(x_i))^2} \qquad\qquad (6.5.14)$$

$$\beta_k = \frac{\displaystyle\sum_{i=0}^{n} w_i (u_k(x_i))^2}{\displaystyle\sum_{i=0}^{n} w_i (u_{k-1}(x_i))^2} \qquad\qquad (6.5.15)$$

The order of the computations is to first compute α_1 and the values of u_1, then β_1, α_2 and the values of u_2, then β_2, etc. until all the u's are known. Using Equation 6.5.12 the a_k's can be computed and thus all terms required by Equation 6.5.8 are known. As an example consider the data in Table 6.5.1.

Point	Y	x
1	7.05	0
2	16.94	1
3	31.16	2
4	48.88	3
5	71.31	4
6	96.81	5
7	127.21	6

Table 6.5.1   Data for Orthogonal Polynomial Example

Assuming unit weighting (i.e., $w_i = 1$), since $u_0 = 1$, from Equation 6.5.14 we compute α_1 as follows:

$$\alpha_1 = \frac{\sum_{i=0}^{n} x_i (u_0)^2}{\sum_{i=0}^{n} (u_0)^2} = \frac{21}{7} = 3$$

and therefore from Equation 6.5.13b $u_1 = x - 3$. We next compute β_1 and α_2 using Equations 6.5.15 and 6.5.14:

$$\beta_1 = \frac{\sum_{i=0}^{n} (u_1)^2}{\sum_{i=0}^{n} (u_0)^2} = \frac{9+4+1+0+1+4+9}{7} = \frac{28}{7} = 4$$

$$\alpha_2 = \frac{\sum_{i=0}^{n} x_i (u_1)^2}{\sum_{i=0}^{n} (u_1)^2} = \frac{0+4+2+0+4++20+54}{9+4+1+0+1+4+9} = \frac{84}{28} = 3$$

and therefore from Equation 6.5.13c $u_2 = (x-3)(x-3) - 4 = x^2 - 6x + 5$. In a similar manner we can compute $\beta_2 = 3$ and $\alpha_3 = 3$ and thus $u_3 = (x-3)u_2 - 3(x-3) = x^3 - 9x^2 + 20x - 6$. To use the u_k's to fit the data we next must compute a_0, a_1, a_2 and a_3 using Equation 6.5.12. The details of the calculation are included in Table 6.5.2.

The results in Table 6.5.2 include four different fits to the data:

$$y = a_0 u_0 = 57.05$$

$$y = a_0 u_0 + a_1 u_1 = 57.05 + 20.01(x - 3)$$

$$y = a_0 u_0 + a_1 u_1 + a_2 u_2$$

$$= 57.05 + 20.01(x - 3) + 2.004(x^2 - 6x + 5)$$

$$y = a_0 u_0 + a_1 u_1 + a_2 u_2 + a_3 u_3$$

$$= 57.05 + 20.01 u_1 + 2.004 u_2 + 0.00389 u_3$$

The terms $S / (n$-p-$1)$ are the sums of the squares of the residuals divided by the number of degrees of freedom. Using the goodness-of-fit criterion explained in Section 3.3 we note that the parabolic equation yields the best results because $S /(n$-p-$1)$ is minimized for $p=2$ (i.e., 3 terms). We can convert this equation to the simple form of Equation 6.5.1:

$$y = (57.05 - 3 * 20.01 + 5 * 2.004) + (20.01 - 6 * 2.004)x + 2.004x^2$$

$$y = 7.04 + 7.986x + 2.004x^2$$

i	Y_i	x_i	u_0	u_1	u_2	u_3
1	7.05	0	1	-3	5	-6
2	16.94	1	1	-2	0	6
3	31.16	2	1	-1	-3	6
4	48.88	3	1	0	-4	0
5	71.31	4	1	1	-3	-6
6	96.81	5	1	2	0	-6
7	127.21	6	1	3	5	6
		$\sum Y_i u_k$	393.36	560.37	168.37	0.84
		$\sum u_k^2$	7	28	84	216
		a_k	57.05	20.01	2.004	0.00389
		S	11552.5	337.7	0.198	0.194
		$S/(n$-p-$1)$	1925.4	67.54	0.049	0.065

Table 6.5.2   Fitting Data using Orthogonal Polynomials

Regardless of the value of p the resulting equation derived using orthogonal polynomials can be converted to the simple very appealing polynomial form (i.e., Equation 6.5.1). For difficult linear problems such as the Filippelli problem this technique avoids the numerical pitfalls arising from the direct use of 6.5.1.

6.6 Multi-Dimensional Models

An important feature of general purpose NLR (nonlinear regression) programs is the ability to handle multi-dimensional problems. Throughout the

book the discussion has primarily been about the relationship between a
dependent scalar variable y and an independent scalar variable x. However, there are many problems throughout many fields of science and engineering where either x or y or both are vector variables. To test NLR
programs it is useful to have a few examples of problems of these types.
Unfortunately the nonlinear regression NIST datasets are limited to problems in which x and y are both scalars.

The theory and use of the **GraphPad Prism** program is included in a book
written by H. Motulsky and A. Christopoulos [MO03]. The book can be
downloaded from the GraphPad Software website (www.graphpad.com)
and includes a very nice example of a problem in which the dependent
variable y is a vector. Although **GraphPad Prism** is a general purpose
NLR program, the book emphasizes analysis of biological and pharmaceutical experiments. Using GraphPad terminology, global models are models
in which y is a vector and some of the unknowns are shared between the
separate models for the components of y. A **GraphPad** example relevant
to the pharmaceutical industry is the use of global models to analyze the
dose-response curves of two groups (a treated group and a control group).
The purpose of the experiment is to measure what they call **ec50** (the dose
concentration that gives a response half-way between the minimum and
maximum responses). For this experiment the x variable is the *log* of the
dose, the first component of the y vector is the response of the control
group and the second component is the response of the treated group. The
problem is well documented in their book and data is included so that the
problem can be used as a test dataset for any NLR program.

The experiment was analyzed using **REGRESS** and the results are very
close to the results obtained with **Graphpad Prism**. The equations were
specified as follows:

```
dependent ycont, ytreat;
independent x;
unknown  bottom, top, hillslope, logec50c,
         logec50t;
ycont  = 'bottom+(top-bottom)/
         (1+10^((logec50c-x)*hillslope))'
ytreat = 'bottom+(top-bottom)/
         (1+10^((logec50t-x)*hillslope))'
```

The two components of the y vector are **ycont** and **ytreat**. The unknown
parameters shared by both equations are **bottom, top** and **hill-
slope**. The two remaining unknowns are the logs of **ec50** for the control

and treatment groups (i.e., `logec50c` and `logec50t`). The data is included in Table 6.6.1. The results are seen in Figure 6.6.1. REGRESS required 9 iterations to converge to the solution. The alternative to the global approach for this problem is to treat each curve separately. The reason for treating this problem using a global model is explained in the **Graphpad** document: the resulting accuracies for the values of **ec50** are reduced considerably using global modeling. The number of degrees of freedom for this problem (i.e., n-p) is $10 - 5 = 5$.

Point	x (log dose)	Ycont	Ytreat
1	-7.0	165	124
2	-6.0	284	87
3	-5.0	442	195
4	-4.0	530	288
5	-3.0	573	536

Table 6.6.1 Data for dose-response curve analysis from Graphpad.

REC	Y-INDEX	X	YCONT	SIGYCONT	CALC_VALUE
1	1	-7.00000	165.000	1.00000	152.28039
2	1	-6.00000	284.000	1.00000	271.95980
3	1	-5.00000	442.000	1.00000	455.54116
4	1	-4.00000	530.000	1.00000	549.35957
5	1	-3.00000	573.000	1.00000	573.06096

REC	Y-INDEX	X	YTREAT	SIGYTREAT	CALC_VALUE
1	2	-7.00000	124.000	1.00000	112.35928
2	2	-6.00000	87.000	1.00000	123.13971
3	2	-5.00000	195.000	1.00000	172.89774
4	2	-4.00000	288.000	1.00000	321.78672
5	2	-3.00000	536.000	1.00000	491.61468

PARAMETER	INIT_VALUE	MINIMUM	MAXIMUM	VALUE	SIGMA
BOTTOM	0.00000	Not Spec	Not Spec	109.781	27.807
TOP	1000.00	Not Spec	Not Spec	578.939	34.182
HILLSLOPE	1.00000	Not Spec	Not Spec	0.72458	0.1845
LOGEC50C	-7.00000	Not Spec	Not Spec	-5.61755	0.1963
LOGEC50T	-2.00000	Not Spec	Not Spec	-3.88429	0.1909

```
Variance Reduction:            97.67 (Average)
   VR:        YCONT            99.26
   VR:        YTREAT           96.08
S/(N - P)         :          1181.32
RMS (Y - Ycalc)   :            24.30351 (all data)
       RMS(Y1-Ycalc):          13.15230
       RMS(Y2-Ycalc):          31.75435
```

Figure 6.6.1 Results from REGRESS analysis of data in Table 6.6.1.

A problem that demonstrates modeling with two independent variables was included in my first book [WO67]. This problem was related to a measurement of parameters related to the neutronics of heavy water nuclear reactors. The model was based upon the following equation:

$$y = \frac{(1 + a_1 x_1)(1 + a_2 x_1)}{(1 + a_1 x_2)(1 + a_2 x_2)} \tag{6.6.1}$$

The unknowns a_1 and a_2 must be positive but there is no guarantee that the method of least squares will satisfy this requirement. However, we can force positive values by simply using b^2 in place of a. The modified equation is thus:

$$y = \frac{(1+b_1^2 x_1)(1+b_2^2 x_1)}{(1+b_1^2 x_2)(1+b_2^2 x_2)} \qquad (6.6.2)$$

The two unknowns are now b_1 and b_2 and regardless of the resulting signs of b_1 and b_2, the squared values are always positive. It should be noted that there are four possible solutions: both b_1 and b_2 can be positive or negative. Depending upon the initial guesses for b_1 and b_2, if convergence is achieved, the solution will be close to one of the four possibilities. The data for this problem is included in Table 6.6.2 and the results of the REGRESS analysis are seen in Figure 6.6.2. Note that for this problem since the σ's vary from point to point Equation 2.3.7 must be used to properly weight the data. The initial guesses were $b_1 = 1$ and $b_2 = 10$ and convergence was achieved with 3 iterations.

```
PARAM INIT_VALUE MINIMUM  MAXIMUM     VALUE      SIGMA
  B1   1.00000  Not Spec Not Spec   1.61876    0.22320
  B2  10.00000  Not Spec Not Spec   5.29172    0.34342

  Variance Reduction:          99.32
  S/(N - P)        :            6.98221
  RMS (Y - Ycalc)  :            0.01946
  RMS ((Y-Ycalc)/Sy):           2.62056
```

Figure 6.6.2   Results from REGRESS analysis of data in Table 6.6.2.

Point	Y	σ_y	x_1	σ_{x1}/x_1	x_2	σ_{x2}/x_2
1	0.7500	0.01000	0.0137	0.0056	0.0258	0.0057
2	0.5667	0.00833	0.0137	0.0056	0.0459	0.0065
3	0.4000	0.00620	0.0137	0.0056	0.0741	0.0070
4	0.8750	0.01243	0.0240	0.0086	0.0320	0.0068
5	0.7000	0.01022	0.0240	0.0086	0.0453	0.0057
6	0.5750	0.00863	0.0240	0.0086	0.0640	0.0054
7	0.3800	0.00586	0.0240	0.0086	0.0880	0.0055
8	0.5750	0.00863	0.0260	0.0093	0.0666	0.0122
9	0.2967	0.00777	0.0260	0.0093	0.1343	0.0134
10	0.1550	0.00290	0.0260	0.0093	0.2291	0.0140
11	0.0900	0.00189	0.0260	0.0093	0.3509	0.0143

Table 6.6.2   Modeling Data for Analysis of Equation 6.6.2.

Note that the value of b_1 is measured to $100 * 0.223 / 1.619 = 13.8\%$ accuracy and b_2 is measured to 6.5% accuracy, but what we are really interested in are the values of a_1 and a_2 and their associated σ's. In general if we have v as a function of u we can relate σ_v to σ_u as follows:

$$\sigma_v^2 = \left(\frac{\partial f}{\partial u}\sigma_u\right)^2 \quad \text{where} \quad v = f(u) \tag{6.6.3}$$

For $v = u^2$ from Equation 6.6.3 we get:

$$\sigma_v^2 = (2u\sigma_u)^2 \quad \text{where} \quad v = u^2 \tag{6.6.4}$$

Dividing the equation by $v^2 = u^4$ we end up with the following simple relationship:

$$\left(\frac{\sigma_v}{v}\right)^2 = \left(\frac{2\sigma_u}{u}\right)^2 \quad \text{where} \quad v = u^2 \tag{6.6.5}$$

In other words the relative uncertainty in v is twice as large as that for u. Using Equation 6.6.5 we see that the relative uncertainties of the a's are twice those of the b's. Thus for the problem in Figure 6.6.2, $a_1 = 1.619^2 = 2.621$ and $\sigma_{a1} = 2.621*2*0.138 = 0.723$. Similarly, $a_2 = 27.99$ and $\sigma_{a2} = 3.64$. It is interesting to note that REGRESS can solve this problem directly for the a's by replacing Equation 6.6.1 by the following alternative:

$$y = \frac{(1 + abs(a_1)x_1)(1 + abs(a_2)x_1)}{(1 + abs(a_1)x_2)(1 + abs(a_2)x_2)} \tag{6.6.6}$$

The abs (absolute) operator is a valid REGRESS operator that can be used in any function specification.

6.7 Software Performance

There are many ways to measure the performance of NLR (nonlinear regression) programs but for most problems the only relevant measure is the ability to converge to a solution for difficult problems. The NIST datasets are very useful for testing the ability of NLR programs to converge and this subject was considered in Sections 6.3 and 6.4. However, there are

some problems where software performance metrics other than convergence are important. In particular, problems in which the amount of data
is large, the time required to converge to a solution may become important.
Another area where time is important is for calculations embedded within
real time systems (e.g., anti-missile missile systems). When decisions
must be made within a fraction of a second, if an NLR calculation is part
of the decision making process, it is important to make the calculation as
fast as possible. For real time applications general purpose NLR software
would never be used. The calculation would be programmed to optimize
speed for the particular system and mathematical model under consideration.

Since time is dependent upon hardware, one would prefer measures that
are hardware independent. In this section some useful measures of performance (other than the ability to converge) are discussed. The total time
that a program requires to achieve convergence for a particular program
and a particular computer is approximately the following:

$$Converge\_Time = Num\_Iterations * Avg\_Time\_per\_Iter \qquad (6.7.1)$$

The number of iterations required to achieve convergence is of course
problem dependent but it can be used as a measure of performance when
used for comparisons with common data sets such as the NIST datasets.
The average time per iteration is of course computer dependent, but the effect of the computer is only a multiplicative speed factor:

$$Avg\_Time\_per\_Iter = Speed\_Factor * Avg\_Calcs\_per\_Iter \qquad (6.7.2)$$

For traditional algorithms such as Gauss-Newton (GN) or Levenberg-
Marquardt (LM) or some sort of combination, the average number of calculations per iteration can be broken down into 2 major terms:

$$Avg\_Calcs\_per\_Iteration = Avg\_CA\_Calcs + Avg\_S\_Calcs \qquad (6.7.3)$$

The first term is a measure of the effort to compute the C matrix and then
the A vector times the average number of times this operation is performed
per iteration. The second term is a measure of the effort to compute the
weighted sum-of-squares S times the average number of times this operation is performed per iteration. Both terms are proportional to n, the number of data points. The first term also has a component that is proportional
to p^3 (the complexity of solving p simultaneous equations).

These equations are meaningless for people evaluating existing software as the actual numbers for a given problem are usually unavailable to the normal user. However, for those interested in developing software for performing NLR analyses for problems with important speed requirements, these equations give some indication where one should concentrate the effort at achieving speed.

For stochastic algorithms, these equations are not applicable. The concept of iterations is not really relevant. The entire calculation becomes essentially a series of calculations of S. Whether or not this results in a faster overall computation is not obvious and clearly the speed of such algorithms is problem dependent.

6.8 The REGRESS Program

Throughout the book results for a number of examples have been obtained using the REGRESS program. The reason why I have chosen REGRESS is quite simple: I wrote it. The program can be downloaded from: www.technion.ac.il/wolberg. The history of the development of this program goes back to my early career when I was in charge of designing a sub-critical heavy water nuclear reactor facility. One of the experiments that we planned to run on the facility involved a nonlinear regression based upon Equation 6.6.2. In the 1960's commercial software was rare so we had no choice other than writing our own programs. It became quite apparent that I could generalize the software to do functions other than Equation 6.6.2. All that had to be done was to supply a function to compute $f(x)$ and another function to compute the required derivatives. We would then link these functions to the software and could thus reuse the basic program with any desired function. At the time we called the program ANALYZER.

In the early 1970's I discovered a language called FORMAC that could be used for symbolic manipulation of equations. FORMAC was compatible with FORTRAN and I used FORTRAN and FORMAC to write a program similar to ANALYZER and I called the new program REGRESS. The REGRESS program accepted equations as input quantities. Using FORMAC, the program automatically generated equations for the derivatives and created FORTRAN subroutines that could then be used to perform the nonlinear regression (NLR). All these steps, including compilation and link-editing of the subroutines, were performed automatically without any user intervention. The REGRESS program became a commercial product on the NCSS time-sharing network and I had the opportunity to work with a number of NCSS clients and learned about many different applications of NLR.

In the mid 1970's I realized that with languages that support recursive programming, I could avoid the need to externally compile subroutines. Recursion is the ability to call a subroutine from within itself. Using recursion, it became a doable task to write a routine to symbolically differentiate functions. Using PL/1 I rewrote REGRESS and added many new features that I realized were desirable from conversations with a number of users of REGRESS. I've returned to the REGRESS program on many occasions since the original version. In the 1980's I started teaching a graduate course called Design and Analysis of Experiments and I supplied REGRESS to the students. Many of the students were doing experimental work as part of their graduate research and the feedback from their experiences with REGRESS stimulated a number of interesting developments. In the early 1990's I rewrote REGRESS in the C language. Through the many version changes REGRESS has evolved over the years and is still evolving.

The REGRESS program lacks some features that are included in other general NLR programs. Some students who have recently used REGRESS have suggested that the program should have a GUI (Graphic User Interface) front end. Such a GUI would give REGRESS the look and feel of a modern program. Personally I have my doubts that this will make the program appreciably more user-friendly and have so far resisted creating such an interface. A more serious problem with REGRESS was the need to create data files in a format that the program could understand. Many users of the program gather data that ends up in an Excel Spread Sheet. The problem for such users was how to get the data into REGRESS. It turned out that the solution was quite simple: Excel allows users to create text files. A feature was added to accept Excel text files. Another important issue was the creation of graphic output. One of the features of REGRESS is that the entire interactive session is saved as a text file. The current method for obtaining graphics output is to extract the output data from the text file and then input it into a program such as Excel that supports graphics. Since this turns out to be a relatively painless process, the need for REGRESS to generate graphic output is not a pressing issue.

The REGRESS program includes some features that are generally not included in other NLR programs. The most important feature in REGRESS that distinguishes it from other general purpose NLR programs is the Prediction Analysis (experimental design) feature described in Chapter 5. Another important feature that I have not seen in other general purpose NLR programs is the *int* operator. This is an operator that allows the user to model initial value nonlinear integral equations. For example consider the following set of two equations:

$$y_1 = a_1 \int_0^x y_2 dx + a_2$$

$$y_2 = a_3 \int_0^x y_1 dx + a_4$$

(6.8.1)

These highly nonlinear and recursive equations can be modeled in REGRESS as follows:

y1 = '*a1* * int(*y2*, 0, x) + *a2*'
y2 = '*a3* * int(*y1*, 0, x) + *a4*'

This model is recursive in the sense that *y1* is a function of *y2* and *y2* is a function of *y1*. Not all general purpose NLR programs support recursive models. The user supplies values of *x*, *y1* and *y2* for *n* data points and the program computes the least squares values of the a_k's.

Another desirable REGRESS feature is a simple method for testing the resulting model on data that was not used to obtain the model. In REGRESS the user invokes this feature by specifying a parameter called NEVL (number of evaluation points). Figure 6.8.1 includes some of the REGRESS output for a problem based upon Equation 6.8.1 in which the number of data records for modeling was 8 and for evaluation was 7. Each data record included values of *x*, *y1* and *y2* (i.e., a total of 16 modeling and 14 evaluation values of *y*). The program required 15 iterations to converge.

```
Function Y1:    A1 * INT(Y2,0,X) + A2
Function Y2:    A3 * INT(Y1,0,X) + A4
```

K	A0(K)	AMIN(K)	AMAX(K)	A(K)	SIGA(K)
1	0.50000	Not Spec	Not Spec	1.00493	0.00409
2	1.00000	Not Spec	Not Spec	2.00614	0.00459
3	0.00000	Not Spec	Not Spec	-0.24902	0.00079
4	-1.00000	Not Spec	Not Spec	-3.99645	0.00663

```
Evaluation of Model for Set 1:
   Number of points in evaluation data set:      14
   Variance Reduction (Average)                 100.00
            VR:           Y1                    100.00
            VR:           Y2                    100.00
   RMS (Y - Ycalc)       (all data)              0.01619
            RMS (Y-Yc) - Y1                       0.02237
            RMS (Y-Yc)/Sy) - Y1                   0.00755
            RMS (Y-Yc) - Y2                       0.00488
            RMS (Y-Yc)/Sy) - Y2                   0.00220
   Fraction Y_eval positive             :    0.214
   Fraction Y_calc positive             :    0.214
   Fraction Same Sign                   :    1.000
```

Data Set	Variable	Min	Max	Average	Std_dev
Modeling	X1	0.0100	6.2832	1.6970	2.3504
Modeling	Y1	-7.9282	2.0000	-1.2393	3.7499
Modeling	Y2	-4.1189	4.0000	-2.2600	3.1043
Evaluate	X1	0.1500	5.2360	1.6035	1.8876
Evaluate	Y1	-8.0000	1.3900	-2.1940	3.4524
Evaluate	Y2	-4.1169	2.9641	-2.6260	2.7180

Figure 6.8.1 Recursion, the *int* operator & evaluation points

Chapter 7 KERNEL REGRESSION

7.1 Introduction

Kernel regression is one class of data modeling methods that fall within
the broader category of *smoothing* methods. The method of least squares
is used within a kernel regression analysis to fit the data within the regions
of interest. The general purpose of smoothing is to find a line or surface
which exhibits the general behavior of a dependent variable (lets call it y)
as a function of one or more independent variables. No attempt is made to
find a single mathematical model for y. If there is only one independent
variable, then the resulting *smoothing* is a line. If the number of independent variables is greater than one, the *smoothing* is a surface. Smoothing
methods that are based upon a mathematical equation to represent the line
or surface are called parametric methods. The method of least squares is
one such parametric method. On the other hand, data driven methods that
only smooth the data without trying to find a single mathematical equation
are called nonparametric methods. An excellent review of nonparmetric
methods is included *The Elements of Statistical Learning* by Hastie,
Tabshirani and Friedman [HA01]. Kernel regression is a nonparametric
smoothing method for data modeling.

The distinguishing feature of kernel regression methods is the use of a *kernel* to determine a weight given to each data point when computing the
smoothed value at any other point on the surface. There are many ways to
choose a kernel. Wolfgang Hardle reviews the relevant literature in his
book on this subject [HA90]. Another overview of the subject by A. Ullah
and H. D. Vinod is included in Chapter 4 of the Handbook of Statistics
Volume 11 [UL93]. Usually the kernel includes one free parameter that
may be adjusted to obtain the "best" fit to the data.

Even though kernel regression utilizes the method of least squares, it is an
alternative technique to standard least squares modeling. It is particularly

useful for problems in which there is no basis for selecting a mathematical model. All that one requires is some method for using the available data for making predictions regarding the dependent variable as a function of the independent variable or variables. Kernel regression is a method for using the data to define a surface that can then be used to obtain estimates of y and the uncertainty associated with y for desired combinations of the x_j's.

Typically kernel regression is applied to multi-dimensional modeling in which there are several independent variables. One application area is econometric modeling [WO00]. Econometric problems are characterized by time series data in which the analyst attempts to use the historical data to make future predictions. The underlying assumption is that historical behavior has relevance regarding the future. Another area in which kernel regression is quite useful is for some medical problems. Medical problems rarely involve time series data. A typical example might involve an attempt to develop a model to predict the probability of contracting a particular disease as a function of personal and environmental variables.

Although the method is usually used for multi-dimensional modeling, to explain the problem consider the data in Figure 7.1.1. The dependent variable y is a function of a single variable x but we have no theoretical basis for suggesting a mathematical model. We also have no basis for suggesting a model for the uncertainties σ_y associated with each of the data points. We would like to have some $f(x)$ that we can use to make predictions for any value of y within the range of observable values (i.e., from $x = 2.7$ to 11.7) and we also require some estimate of the accuracy of the predictions (i.e., σ_f). What we can see from the data is that noise (i.e., σ_y) increases with x, but how can we include this in our model? To weight the data properly we would have to know how σ_y is related to x. Only then would the resulting values of σ_f be meaningful. For this simple one-dimensional problem, we could suggest a relationship to estimate σ_y as a function of x but for multi-dimensional problems this becomes more difficult. This whole issue is avoided when the modeling method is kernel regression.

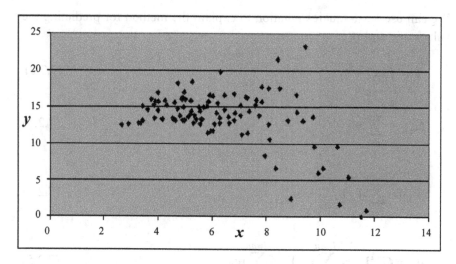

Figure 7.1.1   Observations of *y* versus *x*

7.2 Kernel Regression Order Zero

The simplest kernel regression method is what is called the *Order Zero* method [WO00]. The term order zero is used because a zero-order polynomial is the local fitting function regardless of the dimensionality of the model. A zero order polynomial is just a constant. To make a prediction at any point one first finds the **nn** nearest neighbors of the point (where **nn** is a user specified constant). The weighted average of the values of *y* for these nearest neighbors is then used to predict the value at the point of interest. The uncertainty at this point can also be estimated. The first step is to select a kernel for weighting the nearest neighbor points. Usually the exponential kernel is used:

$$ w_{ij} = exp\!\left(- kd_{ij}^{2}\right) \tag{7.2.1} $$

In this equation the index *i* represents the point at which a value of *y* is to be estimated. The index *j* represents a point within the nearest neighbor set. The term d_{ij} is the distance between the two points. The square of d_{ij} is used for computational convenience since it is always positive. For multi-dimensional models, if the scales of the different dimensions are very different, the usual procedure is to scale all the dimensions so that their ranges are equivalent and thus all values of x_j are in the range 0 to 1.

We can use least-squares notation to explain the method for predicting the values of y and σ_f:

$$y = f(x) = a_1 \tag{7.2.2}$$

$$C_{11} = \sum_{j=1}^{j=nn} w_{ij} \tag{7.2.3}$$

$$V_1 = \sum_{j=1}^{j=nn} w_{ij} Y_j \tag{7.2.4}$$

$$a_1 = C_{11}^{-1} V_1 = \sum_{j=1}^{j=nn} w_{ij} Y_j \Big/ \sum_{j=1}^{j=nn} w_{ij} \tag{7.2.5}$$

Note that i is the index of the point at which an estimate is to be made and is not one of the nn nearest neighbors. From Equation 2.6.11:

$$\sigma_f^2 = \frac{S}{nn-1}\left(\frac{\partial f}{\partial a_1}\frac{\partial f}{\partial a_1} C_{11}^{-1}\right) = \frac{S}{nn-1} C_{11}^{-1}$$

$$= \frac{1}{nn-1} \frac{\displaystyle\sum_{j=1}^{j=nn} w_{ij}(Y_j - a_1)^2}{\displaystyle\sum_{j=1}^{j=nn} w_{ij}} \tag{7.2.6}$$

As an example, consider the data in Table 7.2.1. The first 4 points are used as the nearest neighbors for making predictions on the next 3 points. We call the first 4 points the *learning* points and the next 3 points the *test* points. The model is two dimensional (i.e., $y = f(x_1, x_2)$). For the kernel we will try 3 different values of k ($k = 0$, 1, and 2). A summary of the results is shown in Table 7.2.2. Note that the calculated value of y is the value of a_1 (computed using Equation 7.2.5) and is equal to 3.50 for all 3 test points for the case $k = 0$. From Equation 7.2.1 we see that for this case all the weights are equal so *ycalc* for any point using only learning points 1 to 4 is their average value (i.e., 3.50). Similarly, for point 6 the distance to each of the 4 learning points is the same, so the weights are equal regardless of the value of k and the value of *ycalc* is 3.50.

Point	x_1	x_2	Y
1	0.0	0.0	4.00
2	2.0	0.0	5.00
3	2.0	2.0	3.00
4	0.0	2.0	2.00
5	0.5	0.5	3.75
6	1.0	1.0	3.50
7	1.5	1.5	3.25

Table 7.2.1   Data for Kernel Regression Calculation.
Points 1 to 4 are learning points and 5 to 7 are test points.

Point	Y	ycalc (k=0)	ycalc (k=1)	ycalc (k=2)
5	3.75	3.50	3.88	3.98
6	3.50	3.50	3.50	3.50
7	3.25	3.50	3.12	3.02

Table 7.2.2   Calculated Values of *y* for 3 different values of *k*.

The calculation of *y* for Point 5 for $k = 1$ is shown in Table 7.2.3. The squared distanced are computed as follows:

$$d_{ij}^2 = (x_{1i} - x_{1j})^2 + (x_{2i} - x_{2j})^2 .$$

The value of *ycalc* is 3.034 / 0.7818 = 3.88 as seen in Table 7.2.2. The calculations of σ_f are summarized in Table 7.2.4. Note that for $k = 0$ all the values are equal since all the weights are equal and the values of *ycalc* (i.e., a_1) all equal 3.50. From Equation 7.2.6 :

$$\sigma_f^2 = \frac{1}{3} \frac{(4-3.5)^2 + (5-3.5)^2 + (3-3.5)^2 + (2-3.5)^2}{4}$$

$$= \frac{5}{12} = 0.6455^2$$

Learning Point j	d_{5j}^2	w_{5j}	Y_j	$w_{5j} Y_j$
1	0.5	0.6065	4.0	2.4260
2	2.5	0.0821	5.0	0.4105
3	4.5	0.0111	3.0	0.0333
4	2.5	0.0821	2.0	0.1642
Sum		0.7818		3.0340

Table 7.2.3   Details for Calculation of *y* for Point 5 and *k* = 1.

Test Point	Y	$\sigma_f(k=0)$	$\sigma_f(k=1)$	$\sigma_f(k=2)$
5	3.75	0.6455	0.4183	0.1716
6	3.50	0.6455	0.6455	0.6455
7	3.25	0.6455	0.4183	0.1716

Table 7.2.4   Calculated Values of σ_f for 3 different values of *k*.

7.3 Kernel Regression Order One

The *Order One* method of kernel regression is based upon a first-order polynomial as the local fitting function regardless of the dimensionality of the model. To make a prediction at any point one first finds the **nn** nearest neighbors of the point (where **nn** is a user specified constant). The algorithm used for order-one is similar to the order zero algorithm. Equation 7.2.1 can be used to specify the kernel but Equation 7.2.2 must be expanded to specify a first order polynomial:

$$y = f(\mathbf{X}) = \sum_{k=1}^{k=d} a_k x_k + a_{d+1} \qquad (7.3.1)$$

In this equation $\mathbf{X}$ is a *d* dimensional vector of the independent variables x_1 thru x_d. There are *d*+1 unknown values of a_k and therefore **nn** must be greater than *d*+1 to permit a least squares local fit. For the order zero algorithm, the *C* matrix and *V* vector contains only a single term. For order one, Equation 7.2.3 is replaced by a matrix in which the terms are specified using Equation 2.4.14 and Equation 7.2.4 is replaced by a vector in which the terms are specified using Equation 2.4.15 :

$$C_{jk} = \sum_{i=1}^{i=n} w_i \frac{\partial f}{\partial a_j} \frac{\partial f}{\partial a_k} \quad j = 1 \text{ to } d+1, \ k = 1 \text{ to } d+1 \qquad (7.3.2)$$

$$V_k = \sum_{i=1}^{i=n} w_i Y_i \frac{\partial f}{\partial a_k} \qquad k = 1 \text{ to } d+1 \qquad (7.3.3)$$

From Equation 7.3.1 the partial derivatives are:

$$\frac{\partial f}{\partial a_k} = x_k \qquad k = 1 \text{ to } d \qquad (7.3.4)$$

$$\frac{\partial f}{\partial a_{d+1}} = 1 \qquad (7.3.5)$$

The vector A is computed by solving the matrix equation $CA = V$. The values of σ_f are computed as follows:

$$\sigma_f^2 = \frac{S}{nn - d - 1} \sum_{k=1}^{k=d+1} \sum_{j=1}^{j=d+1} \frac{\partial f}{\partial a_j} \frac{\partial f}{\partial a_k} C_{jk}^{-1} \qquad (7.3.6)$$

The weighted sum of the squares S is computed as follows:

$$S = \sum_{j=1}^{j=nn} w_{ij} (Y_j - ycalc_j)^2$$
$$= \sum_{j=1}^{j=nn} w_{ij} \left(Y_j - \sum_{k=1}^{k=d} a_k x_{kj} - a_{d+1} \right)^2 \qquad (7.3.7)$$

Note that i is the index of the point at which an estimate is to be made and is not one of the nn nearest neighbors.

As an example, once again consider the data in Table 7.2.1. The first 4 points are used as the nearest neighbors for making predictions on the next 3 points. We call the first 4 points the *learning* points and the next 3

points the *test* points. The model is two dimensional (i.e., $y = f(x_1, x_2)$).
For the kernel we will try 3 different values of k ($k = 0$, 1, and 2). A summary of the results is shown in Table 7.3.1. Note that the values of *ycalc*
are all exactly equal to y for all the values of k. The explanation for this
somewhat surprising result is that the 4 learning points fall exactly on the
plane $y = 0.5x_1 - x_2 + 4$. Thus regardless of how the points are weighted,
the calculated test points will fall on this plane.

Point	Y	ycalc (k=0)	ycalc (k=1)	ycalc (k=2)
5	3.75	3.75	3.75	3.75
6	3.50	3.50	3.50	3.50
7	3.25	3.25	3.25	3.25

**Table 7.3.1   Values of *ycalc* for 3 different values of k. Data from
Table 7.2.1.**

To make the calculation more interesting let us change y for the first learning data point from 4 to 5. Table 7.2.1 is thus replaced by Table 7.3.2.
The results are summarized in Table 7.3.3. The calculations of σ_f are
summarized in Table 7.3.4. For all 3 values of k the value of *ycalc* = 3.75
for Point 6. Since all 4 of the learning points are equidistant from Point 6,
the weights are all equal regardless of the value of k. The same plane is
computed for all these cases (i.e., $y = 0.25x_1 - 1.25x_2 + 4.75$) and the value
at $x_1 = 1$ and $x_2 = 1$ is 3.75. The values for σ_f for Point 6 are all 0.25. This
value is obtained from Equation 7.3.7:

$$\sigma_f^2 = \frac{S}{4-2-1}(x_1^2 C_{11}^{-1} + x_2^2 C_{22}^{-1} + C_{33}^{-1}$$
$$+ 2(x_1 x_2 C_{12}^{-1} + x_1 C_{13}^{-1} + x_2 C_{23}^{-1}))$$
$$\sigma_f^2 = S(C_{11}^{-1} + C_{22}^{-1} + C_{33}^{-1} + 2(C_{12}^{-1} + C_{13}^{-1} + C_{23}^{-1}))$$

The value of S is 0.25 and the matrices C and C^{-1} are:

$$C = \begin{bmatrix} 8 & 4 & 4 \\ 4 & 8 & 4 \\ 4 & 4 & 4 \end{bmatrix} \quad \text{and} \quad C^{-1} = \begin{bmatrix} 0.25 & 0 & -0.25 \\ 0 & 0.25 & -0.25 \\ -0.25 & -0.25 & 0.75 \end{bmatrix}$$

It should be emphasized, that although the same four points are used to define a plane for this two dimensional problem, for the cases where the points are not equally weighted (i.e., $k = 1$ and 2), the planes are different for the different test points. The different planes are listed in Table 7.3.5. Note once again that the planes for Point 6 are the same regardless of the value of k because this point is equidistant from the 4 learning points.

Point	x_1	x_2	y
1	0.0	0.0	5.00
2	2.0	0.0	5.00
3	2.0	2.0	3.00
4	0.0	2.0	2.00
5	0.5	0.5	3.75
6	1.0	1.0	3.50
7	1.5	1.5	3.25

Table 7.3.2   Data for Kernel Regression Calculation.

Point	Y	ycalc (k=0)	ycalc (k=1)	Ycalc (k=2)
5	3.75	4.25	4.30	4.26
6	3.50	3.75	3.75	3.75
7	3.25	3.25	3.30	3.26

Table 7.3.3   Values of ycalc for 3 different values of k. Data from Table 7.3.2.

Point	Y	σ_f (k=0)	σ_f (k=1)	σ_f (k=2)
5	3.75	0.3062	0.1209	0.0470
6	3.50	0.2500	0.2500	0.2500
7	3.25	0.3062	0.3062	0.0470

Table 7.3.4   Calculated Values of σ_f for 3 different values of k.

Pt	x_1	x_2	Plane (k=1)	Plane (k=2)
5	0.5	0.5	0.0596x_1- 1.4404x_2+4.9558	0.0090x_1- 1.4910x_2+4.9997
6	1.0	1.0	0.2500x_1- 1.2500x_2+4.7500	0.2500x_1- 1.2500x_2+4.7500
7	1.5	1.5	0.4404x_1- 1.0596x_2+4.2242	0.4910x_1- 1.0090x_2+4.0356

Table 7.3.5   Planes for each Test Point ($k = 1$ and 2).

7.4 Kernel Regression Order Two

The *Order Two* method of kernel regression is based upon a second-order polynomial as the local fitting function regardless of the dimensionality of the model. This algorithm is the next logical step after the order-zero and order-one algorithms discussed in the previous sections. Clearly we can continue to propose higher and higher order algorithms but this is not a reasonable approach to modeling. The number of constants required by the local fitting function increases dramatically as the number of independent variables increases. The local complete second order fitting function for a *d* dimensional space is:

$$y = f(\mathbf{X}) = a_1 + \sum_{j=1}^{d} a_{j+1} x_j + \sum_{j=1}^{d} \sum_{k=j}^{d} b_{jk} x_j x_k \qquad (7.4.1)$$

In this equation $\mathbf{X}$ is a *d* dimensional vector. The number of constants *p* in this equation is $1 + d + d(d+1) / 2$. For example for $d = 1$ this equation reduces to:

$$y = a_1 + a_2 x_1 + b_{11} x_1^2 \qquad (7.4.2)$$

For $p = 2$ this equation reduces to:

$$y = a_1 + a_2 x_1 + a_3 x_2 + b_{11} x_1^2 + b_{12} x_1 x_2 + b_{22} x_2^2 \qquad (7.4.3)$$

The number of constants in the fitting function is listed in Table 7.4.1 for the orders zero, one and two algorithms as a function of *d*.

d	Order-zero	Order-one	Order-two
1	1	2	3
2	1	3	6
3	1	4	10
4	1	5	15
5	1	6	21
6	1	7	28

Table 7.4.1   Number of Constants p in Fitting Function as a Function of the number of independent variables d.

As with the order zero and one algorithms, to make a prediction at any point one first finds the **nn** nearest neighbors of the point (where **nn** is a user specified constant). From Table 7.4.1 we see that as the number of independent variables d increases, the value of **nn** must also be increased to maintain the same number of degrees of freedom (i.e., **nn** $-$ **p**) which of course must be greater than one. For noisy data, order-two is not particularly useful for larger values of d unless **nn** is large. The large number of constants in the fitting function tends towards a fit that accommodates the noise in the data.

The A vector is the vector of the unknown constants. The vector A is computed by solving the matrix equation $CA = V$. The C matrix and V vector are computed in the usual manner. For example, for $d = 1$ the A vector is $[a_1 \; a_2 \; b_{11}]^t$ and C and V are:

$$C = \begin{bmatrix} \sum w_i & \sum w_i x_{1i} & \sum w_i x_{1i}^2 \\ \sum w_i x_{1i} & \sum w_i x_{1i}^2 & \sum w_i x_{1i}^3 \\ \sum w_i x_{1i}^2 & \sum w_i x_{1i}^3 & \sum w_i x_{1i}^4 \end{bmatrix} \qquad (7.4.4)$$

$$V = \begin{bmatrix} \sum w_i Y_i \\ \sum w_i x_{1i} Y_i \\ \sum w_i x_{1i}^2 Y_i \end{bmatrix} \qquad (7.4.5)$$

The summations in these equations are over all **nn** points (i.e., $i = 1$ to **nn**). For $d = 1$ the values of σ_f are computed as follows:

$$\sigma_f^2 = \frac{S(C_{11}^{-1} + x_1^2 C_{22}^{-1} + x_1^4 C_{33}^{-1} + 2(x_1 C_{12}^{-1} + x_1^2 C_{13}^{-1} + x_1^3 C_{23}^{-1}))}{nn - 3} \tag{7.4.6}$$

The weighted sum of the squares S is computed as follows:

$$S = \sum_{j=1}^{j=nn} w_{ij}(Y_j - ycalc_j)^2$$

$$= \sum_{j=1}^{j=nn} w_{ij}(Y_j - a_1 - a_2 x_{1j} - a_3 x_{1j}^2)^2 \tag{7.4.7}$$

Note that i is the index of the point at which an estimate is to be made and is not one of the nn nearest neighbors.

As an example, consider the data in Table 7.4.2. Let us use the 4 nearest neighbors to predict the value of y for the all the points.

Point	x_1	Y	Nearest Neighbors
1	0.0	10.00	2,3,4,5
2	0.5	6.06	1,3,4,5
3	1.0	3.68	1,2,4,5
4	1.5	2.23	2,3,4,6
5	2.0	1.35	3,4,6,7
6	2.5	0.82	3,4,5,7
7	3.0	0.50	3,4,5,6

Table 7.4.2   Data for Kernel Regression Order-Two Demonstration

For the kernel we will try 3 different values of k (k = 0, 1, and 2). Note that a value of $k = 0$ results in all points being equally weighted. A summary of the results is shown in Table 7.4.3. For this particular example, we see that based upon the root mean square error, the best results are obtained for $k = 2$. The rms error is computed as follows:

$$rms\_error = \sqrt{\sum(Y - ycalc)^2 / n} \tag{7.4.8}$$

The summation in this equation is over all the points. We can use the rms error as a criterion for selecting the best value of k. Increasing k in increments of one, the value of the rms error is a minimum equal to 0.262 at $k = 12$. The comparison of values of the rms error is interesting, but are the

results significant? Comparing the values at $k = 0$ and $k = 12$ the ratio is
0.418 / 0.262 = 1.59. The F test described in Section 3.3 can be used to
test significance. For this example, the F statistic is :

$$F = \frac{\left(rms\_error_{k=0}\right)^2}{\left(rms\_error_{k=12}\right)^2} = 1.59^2 = 2.53$$

The number of degrees of freedom in both the numerator and denominator
is $n = 7$ and the value for 1% significance is 6.99 so the observed value of
F is very far from this 1% significance level. It is not even significant at
the 10% level which is 2.78 [ME92]. The problem here is that we just have
too few data points to make a definitive statement about an optimum value
of k.

Point	Y	ycalc (k=0)	ycalc (k=1)	ycalc (k=2)
1	10.00	9.100	9.247	9.338
2	6.06	6.527	6.397	6.313
3	3.68	3.955	3.876	3.826
4	2.23	2.352	2.334	2.317
5	1.35	1.337	1.337	1.337
6	0.82	0.713	0.735	0.756
7	0.50	0.805	0.715	0.660
Rms-error		0.418	0.335	0.283

Table 7.4.3   Values of *ycalc* for 3 different values of *k*. Data from
Table 7.4.2.

7.5 Nearest Neighbor Searching

In the previous sections we described three kernel regression algorithms:
Orders Zero, One and Two. The first step for each of these algorithms is
to specify *nn*, the number of nearest neighbors used to compute the local
least squares value for the dependent variable. When the number of independent variables *d* is one, the search for nearest neighbors is trivial. All
one need do is sort the data based upon the values of the independent variable. For *d* greater than one the problem is more complicated. However,
if the number of data points is not excessively large, all one needs to do is
to compute the distance to each point and then sort by distance. To avoid
problems associated with the sign of the distance, the usual procedure is to

use distance squared. Another problem that arises when d is greater than one is related to the scale of each of the independent variables. If these scales are very different, then the distances in the different directions must be normalized in some manner. The most obvious method of normalization is to specify the minimum value of x as zero, the maximum value as one and every other point as:

$$x_{normalized} = \frac{x - x_{min}}{x_{max} - x_{min}} \tag{7.5.1}$$

Kernel regression is a very attractive modeling method for many applications in which the relationship between the dependent and independent variables is complicated and unknown. For example, kernel regression has been applied to econometric and financial market modeling [WO00]. However, if n, the number of data points is large, then the nearest neighbor search requires $n * (n-1)$ calculations of distance (or distance squared). Note that using distance squared avoids the need to perform a square root operation in every distance calculation. For each point we would have to compute the distance to every other point. We could, of course, save all the distances (or distances squared) in a gigantic matrix and that would save half of the calculations. However, we see that the time devoted to computing distances increases as $O(n^2)$. When n is large, a popular modeling strategy is to divide the data into learning and test data where $nlrn + ntst = n$. This reduces the number of calculations of distance to $nlrn * ntst$ but if both are proportional to n the number of calculations still increases with n^2.

If one is willing to accept an approximate nearest neighbor search, then the time required to find the nearest neighbors can be reduced dramatically [WO00]. If nn is large compared to one, then it is not really necessary to get the exact nn nearest neighbors. For example if nn is 50, if we miss the 7th, 23rd and 34th nearest neighbors and instead use the 51st, 52nd and 53rd nearest neighbors, the effect upon the local least squares solution should be small. The key to an approximate nearest neighbor search is to organize the data into a suitable data structure that can be used over and over again for each search. If the data has been subdivided into learning and test data, then the learning data is first inserted into the data structure and this data structure is used to find the approximate nearest neighbors for each of the test data points. If all n points are treated as test points, then all the points would be included in the data structure and $nn+1$ points would be found for each point (including the point itself). For example, if $n = 100$ and

$nn = 5$ and we are looking for the 5 nearest neighbors to points 27, a search for 6 points might turn up points 14, 23, 27, 38, 84 and 92. After rejecting point 27 we would be left with the 5 approximate nearest neighbors: 14, 23, 38, 84 and 92.

The nearest neighbor problem (often called the K nearest neighbor problem) has been studied by many researchers [e.g., HA90, SK97, SM00]. A library called "ANN: Library for Approximate Nearest Neighbor Searching" can be downloaded from the internet [MO98]. In my last book I described a program called FKR that uses the *p-tree* approach to nearest neighbor searching to perform kernel regression analyses based upon either the Order Zero, One or Two KR algorithm [WO00]. The *p-tree* is a full binary tree of height h and thus contains 2^h leaves. Each leaf of the tree contains information about a region in the independent variable space. If the number of independent variables is d then the height h of the tree must be greater than or equal to d. It is easiest to explain this approach to nearest neighbor searching if we assume that $d = 2$.

In Figures 7.5.1 and 7.5.2 we see 24 points distributed in a *p-tree* of height $h = 3$. There are $2^3 = 8$ cells in this tree. The distances OA and OR are normalized to a value of 1. The tree is constructed by first finding the point on the normalized x_1 axis in which half the data points fall to the left and half fall to the right. If the number of data points n is odd (i.e., $n = 2m+1$), then m will fall on one side and $m+1$ on the other side. In the example we see that the space ADRO is first subdivided into two subspaces ACPO and CDRP. Both of these spaces contain 12 data points. These spaces are then subdivided along the x_2 axis. For example ACPO is subdivided into EGPO and ACGE. Each of these subspaces contains 6 data points. For higher dimensional spaces, this procedure is continued until every direction has been subdivided once. Once the space has been subdivided in every direction, then the next subdivision for each subspace is along the longest normalized direction. For example, cell ACGE is longer in the x_1 than the x_2 direction so it is divided into two subspaces along the x_1 axis (i.e., cells 3 and 4). Cell CDLJ is longer in the x_2 direction so it is divided into two subspaces along the x_2 axis (i.e., cells 7 and 8).

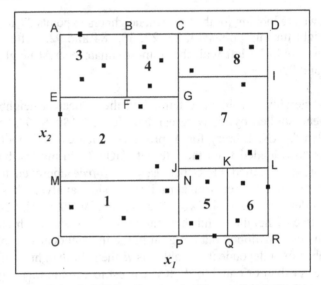

Figure 7.5.1   Cell Distribution in Normalized 2D Space

The *p-tree* data structure includes the dimensions of each of the cells and a list of data points in each of the leaf cells. To find the **nn** approximate nearest neighbors for a given test point, one must first find in which leaf cell the test point resides. To find the test cell, one enters the tree at the root cell and then follows the tree down to the appropriate leaf cell through a series of *if* statements (e.g., if $x_2 > 0.56$ go to *right-son*, else go to *left-son*). The search for nearest neighbors is limited to a maximum of **numcells** cells in which only the test cell and adjacent cells are included in the search. If **numcells** = 1 then the search is limited to the test cell. If **numcells** = 2, then the search is limited to the test cell and the closest adjacent cell. If **numcells** $\geq 2^h$ then the search is performed in the test cell and all adjacent cells. Clearly, the greater the value of **numcells**, the more accurate is the search but the time required to perform the search is greater.

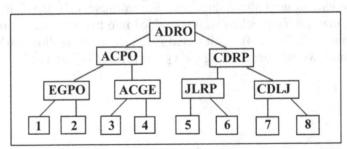

Figure 7.5.2   *p-tree* Representation of Cells in Figure 7.5.1

To understand the approximate nature of this method of locating nearest neighbors, consider Figure 7.5.3. In this figure three test points are considered: one in cell 1, one in cell 2 and one in cell 6. Consider first the test point in cell 1. If **numcells** = 8, and **nn** = 4, then the search would take place only in cells 1, 2 and 5 and the four points that would result from the search would be **a, b, d** and **i** but not **c** (since **c** is not in an adjacent cell). Even if **nn** = 24, the search would only locate 9 learning points (i.e., all the points in cells 1, 2 and 5). The test point in cell 2 presents a different problem. If **numcells** = 3, and **nn** = 2, then the search would be in cells 2, 4 and 7 and points **e** and **f** would be located. However, if **numcells** = 2 then the search would locate **e** and **c** (because the 2nd cell included in the search would be 7 rather than 4). For the test point in cell 6, if **numcells** = 8, and **nn** = 3 then the search would locate the 3 points in cell 6. If **nn** is increased to 4 then point **g** would also be located. If **nn** is increased to 10, then only the 9 points in cells 5, 6 and 7 would be located. Points **a** and **b** are closer to the test point than point **h** but they would not be located because they are not in an adjacent cell.

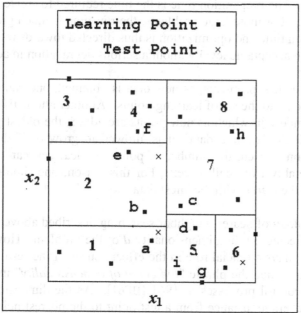

Figure 7.5.3   Data Point Distribution in *p-tree* with *h* = 3 and 24 learning points.

When n is large, there can be many points in each cell. For example, if $n = 10,000$ and $h = 8$, then the average number of points per cell is $10000/2^8$ which is between 39 and 40. One option is to just use all the points in the test cell to make a prediction for the dependent variable of the test point. Furthermore, if all the points in the cell are equally weighted, then the surface for the cell (using Order Zero, One or Two) can be saved and used for every test point falling within the same cell. This procedure allows very rapid predictions but at the price of reduced accuracy. The loss of accuracy is greatest for points near the cell boundaries. To reduce this loss, one can increase **numcell** to 2 and use all the points in the test cell and the closest adjacent cell. Since all points in the two cells are used, there would be no need to perform a nearest neighbor search within this set of learning points. For very large problems (like modeling financial markets) the timing considerations for nearest neighbor searching and then least squares fitting to perform kernel regression analyses becomes crucial. This subject is considered in detail in my last book [WO00]. The time for a complete analysis can be divided into two components: *preparation-time* and *run-time*. The preparation time is the time required to create the *p-tree* data structure. For most large problems, the run-time is much greater than the preparation time and optimization is thus directed towards reducing the run-time to an acceptable level without a serious degradation in accuracy.

For time dependent problems, as new data is obtained, one can choose to add the new data to the set of learning points. Another alternative is to use a moving window in which as new points are added, the oldest points are discarded. If the learning data set is allowed to grow or if the moving window option is used, the number of points per leaf cell can eventually vary considerably from cell to cell. For this reason, one should periodically rebuild the *p-tree* with the latest data.

The methodology of nearest neighbor searching described above is general and is not affected by the dimensionality d of the problem. However, dimensionality plays a crucial role in the effectiveness of the resulting models. Bellman coined the phrase "*the curse of dimensionality*" in his book on adaptive control processes in 1961 [BE61]. As the dimensionality increases, the average distance from a test point to the nearest neighbors increases dramatically. To understand the problem, let us organize n points such that $n/2$ are learning points and $n/2$ are test points and the points are placed within a hypercube of dimension 1 per side. Furthermore the points are placed alternatively at equal distances. For example if we have a single dimension (i.e., $d = 1$), the distance from a test point to the closest learning points would be $1/n$. If $n = 100$ then the distance would be 0.01. In two dimensions (i.e., $d = 2$), there would be $n^{1/2}$ points per side and the distance from a test point to the closest learning points would be $1/n^{1/2}$. For example, for $n = 100$, the layout of the 100 points is shown in Figure 7.5.4. We see that the distance between points is 0.1 (i.e., $1/100^{1/2}$). For $d=3$ the distance increases to 0.215 ((i.e., $1/100^{1/3}$). Distances are summarized in Table 7.5.1 for $n = 100$, 10,000 and 1,000,000 for values of d up to 10. We see that the distance to the closest learning point increases rapidly with increasing d. For example, comparing the distances for $d=10$ and $d=2$ for $n = 10,000$, we see that the ratio is almost 40 (i.e., 0.3981 / 0.01). For $n = 1,000,000$ the ratio is over 250. In other words, for a given value of n, the data becomes increasingly sparse as d increases. In fact, the data density decreases exponentially with increasing d. Thus our search for nn nearest neighbors results in a volume that increases in size exponential with increasing d. For large values of n our nearest neighbors are quite far away thus the basic assumption that the model produces predictions based upon behavior of nearby points becomes suspect. The "nearby points" are not so nearby! This effect is called the *curse of dimensionality*.

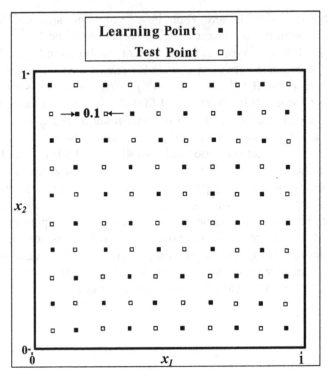

Figure 7.5.4   Layout of 100 data points, 50 learning and 50 test points. Points are equidistant. For each test point the distance to the nearest learning points is 0.1.

d	$n=100$	$n=10,000$	$N=1,000,000$
1	0.0100	0.0001	0.000001
2	0.1000	0.0100	0.0010
3	0.2154	0.0464	0.0100
4	0.3162	0.1000	0.0316
5	0.3981	0.1584	0.0631
6	0.4642	0.2154	0.1000
7	0.5179	0.2682	0.1389
8	0.5623	0.3612	0.1778
9	0.5994	0.3594	0.2154
10	0.6309	0.3981	0.2512

Table 7.5.1   Distance from test points to the nearest learning points in an equidistant grid as a function of dimensionality d and the total number of points n (with $n/2$ learning and $n/2$ test points). *Distance* = $1 / n^{1/d}$.

7.6 Kernel Regression Performance Studies

In this section two studies are considered. Both are based upon a complicated function of four independent variables. Each of the independent variables was generated using a random number generator from zero to one. The values of y for the first case were computed directly from the function. The values of y for the second case are the same as for the first case but a large random noise component was added to the values of y. The equation for y (without noise) is:

$$
\begin{aligned}
y = {} & e^{-2\left(x_1 - 0.25\right)^2} - e^{-2\left(x_1 - 0.75\right)^2} \\
& + e^{-2\left(x_2 - 0.5\right)^2} - e^{-2\left(x_3 - 0.5\right)^2} \\
& + e^{-2\left(x_4 - 0.25\right)^2} - e^{-2\left(x_4 - 0.75\right)^2}
\end{aligned}
\tag{7.6.1}
$$

This equation has multiple peaks and valleys and an average value of zero. The variable y is a function of 4 independent variables. For Case One, 10000 points were generated and 7000 were used as learning points. Predictions were made for each of the remaining 3000 test points and the variance reduction was computed using all the test points:

$$
VR = 100 \, \frac{\displaystyle\sum_{i=1}^{ntst} \left(y_i - ycalc_i\right)^2}{\displaystyle\sum_{i=1}^{ntst} \left(y_i - y_{avg}\right)^2}
\tag{7.6.2}
$$

A *p-tree* of height $h = 8$ was used so the learning points were distributed into 256 approximately equally populated cells. The average number of learning points per cell was 7000 / 256 = 27.3. Nearest neighbor searches were conducted in all adjacent cells to each test cell. The **nn** points located in the searches were weighted using Equation 7.2.1. The calculation was repeated using eight different values of k. The values of k varied from calculation to calculation. The k's were computed using Equation 7.6.3 and eight different values of C (i.e., $C = 1, 2, 4, 8, 16, 32, 64$ and 128) :

$$\frac{1}{C} = w_{min} = exp\left(- kd_{max}^2\right) \tag{7.6.3}$$

In this equation d_{max} is the distance from test point i to the furthest of the nn learning points. For example, if $C = 4$, then the minimum weight is 0.25. All the other $nn - 1$ learning point are weighted between 1 and 0.25. For $C = 1$, k is zero, and all the learning points are equally weighted. Table 7.6.1 includes results using $nn = 20$ and 100 for all three algorithms. Note that all the values of VR are close to 100%. Only results for $C = 1$ and 128 are included in the table, because VR varies smoothly and monotonically between these two values. The results show that there is a slight advantage to using a higher value of C (for all three algorithms) because the higher the value of C the lower the weight for distant points. Also, the results indicate that accuracy is improved as the order of the algorithm increases.

nn	Order	VR (C=1)	VR (C=128)
20	0	99.239	99.497
20	1	99.932	99.969
20	2	99.999	99.999
100	0	97.930	99.203
100	1	99.670	99.912
100	2	99.998	99.999

Table 7.6.1   Values of Variance Reduction for Combinations of nn and Order. ($h = 8$, $nlrn = 7000$, $ntst = 3000$)

In Case 2, Equation 7.6.1 was used to generate values of y but these values were then corrupted with random noise. Defining $rnum(n)$ as a function that generates n random numbers between -1 and 1, the following method was used to generate the values of $y10$:

$$z = rnum(n)$$

$$y10 = y + z\frac{\sigma_y}{\sigma_z}\sqrt{\frac{0.9}{0.1}} \tag{7.6.4}$$

The variable $y10$ is 10% signal (i.e., y) and 90% noise. If independent test data is used, then one expects VR to be less than 10%. Using Equation 7.6.4, 50,000 data points were generated. Thirty-five thousand were used as learning points to make predictions on the remaining fifteen thousand. Results are seen in Table 7.6.2. The values of C in the table are the values for which *VR* is maximized. The results show that the Order 0 algorithm is least sensitive to *nn* and Order 2 is most sensitive. As *nn* increases, Order 0 requires a greater value of *C* to reduce the effect of distant points. For all algorithms, it can be seen that if enough learning points are used to make predictions, then results approach the maximum expected value of *VR* = 10%.

The purpose of these two studies is to demonstrate the power of kernel regression as a modeling tool. For problems in which there is no obvious underlying mathematical model, kernel regression can be used to make predictions for combinations of the independent variables within the range of the learning data. The method can be used for problems in which the data exhibits very little noise as well as for problems in which the data includes considerable noise.

	Order 0	*Order 0*	*Order 1*	*Order 1*	*Order 2*	*Order 2*
nn	*VR*	*C*	*VR*	*C*	*VR*	*C*
50	7.923	1	6.963	1	-11.021	1
100	8.812	1	8.074	1	0.514	1
200	9.379	1	9.096	1	5.063	1
400	9.502	2	9.462	1	7.628	1
800	9.537	8	9.686	1	8.799	1
1600	9.491	128	9.737	4	9.454	1

Table 7.6.2   **Values of *VR* and Best *C* for Combinations of *nn* and Order** . (*h* = 8, *nlrn* = 35000, *ntst* = 15000)

7.7 A Scientific Application

Kernel regression can be applied to problems is many fields of science. I was involved in a study related to the behavior of materials under deformation [TA03]. Clearly, this is not a subject that is of interest to most readers. However, I am including it to illustrate the performance aspects of applying kernel regression to a very compute intensive application.

Typically materials are treated as a continuum but a growing field of study considers the behavior of materials at the atomic level. In many cases, both the finite dimensions of the system as well as the microscopic atomic-scale interactions contribute equally to the overall system response. This makes modeling difficult since continuum tools appropriate to the larger distance scales are unaware of atomic detail and atomistic models are too computationally intensive to treat the system as a whole. One approach is to model such systems using quasicontinuum techniques. Quasicontinuum methods mix continuum and atomistic approaches to modeling. Our study applied a kernel regression Order Two model to simulate the behavior in the atomic region.

The overall strategy is to consider the energetics of the entire system with the aim of finding the configuration in which the stored energy is minimized. Atomistic simulations are performed in the critical regions (for example, near propagating cracks in the material) and continuum methods are used further from these regions. To move smoothly from the atomistic to the continuum regions, as the distance from the critical regions increases, the number of atoms used in the simulation decreases. From an energetics point of view, the atoms are considered as representative of their immediate neighborhood.

In the atomistic regions, what are required are the displacements of the atoms as functions of the positions of the atoms. In three dimensions the position of an atom is denoted as x_1, x_2, x_3 and the displacements are denoted as u_1, u_2, u_3. Crucial to the energetics calculations are the first and second derivatives of the u variables. Each of the u's are modeled separately. For each atom, the first task is to find a set of nn nearest neighbors. Once the nearest neighbors have been located, the method of least squares is used to find the coefficients of the following equation:

$$
\begin{aligned}
u = a_1 x_1 + a_2 x_2 + a_3 x_3 + a_4 x_1^2 + a_5 x_1 x_2 + a_6 x_1 x_3 \\
+ a_7 x_2^2 + a_8 x_2 x_3 + a_9 x_3^2
\end{aligned}
\tag{7.7.1}
$$

In the neighborhood of each atom of interest we use the x's and the u's relative to this atom. In other words, the values of x_1, x_2, x_3 and u for the atom of interest are zero and thus there is no need for a constant in the equation. This equation is a complete second order polynomial in three dimensions. Since this equation includes 9 coefficients, the method of least squares requires a value of nn greater than 9. For each atom there are 9 a_j values for u_1, 9 values for u_2 and 9 values for u_3. The first and second derivatives for each u in each direction can be computed by differentiating the equation. For example, the derivatives in the x_1 direction within the region near the atom of interest are:

$$\frac{\partial u}{\partial x_1} = a_1 + 2a_4 x_1 + a_5 x_2 + a_6 x_3$$

$$\frac{\partial^2 u}{\partial x_1^2} = 2a_4$$

(7.7.2)

Since we are only interested in the derivatives at each atom of interest, the x values are all zero and so the first derivative is just a_1 and the second derivative is just $2a_4$. Similar expression can be determined for the derivatives in the x_2 and x_3 directions. Using the derivatives an expression for total energy can be obtained and then the actual u values can be computed by minimization of the energy equation. This is a non-trivial matter as the number of degrees of freedom is very large. We worked with about 10,000 degrees of freedom (3 times the number of atoms used to represent the system).

An interesting aspect of the study was the validation process used to test the software and measure the effect of the input parameters upon system performance. A test data set was generated and was based upon a random set of x values but u values were based upon a known second order polynomial (in 3 dimensions). Thus computed and actual derivatives could be compared to make sure that the software was behaving properly. (The expected errors should be close to zero and due only to round-off errors.) The input parameters considered in the study included:

1. n : This parameter is the number of data points used in the analysis. All n points were used to build the p-tree (required for nearest neighbor searching) and then the same points were each used individually as test points.

2. *numcells* : This parameter is the number of cells used in the search for nearest neighbors. Each test point falls into a test cell. If *numcells* = 1, then only the test cell is used. However, if *numcells* > 1 then adjacent cells may be used. The value *num_adjacent* is the number of adjacent cells and varies from cell to cell. It is a computed parameter and not an input parameter. If *numcells* > *num_adjacent*, then all adjacent cells are used. If, however, *numcells* < *num_adjacent* then the cells are first sorted on the basis of the distance from their centers to the test point and the closest cells are used up to a total of *numcells*.

3. *numleaves* : This parameter is the number of leaves in the binary tree and must be equal to 2^h where h is an integer and is a measure of the tree height. The number of leaves determines the number of data points included in each cell (*n / numleaves*). If *n / numeaves* is not an integer number than the number of data points per cell is this number rounded up or down by one.

4. *nn* : This parameter is the number of nearest neighbors that are to be located in the search and then used in the least squares analyses. This number is a maximum in the sense that the actual number returned might be less than *nn*. This can happen if the values of *n*, *numleaves*, and *numcells*, result in the number of points under consideration being less than *nn*. For example, if $n = 2560$ and *numleaves* $= 256$, then there will be 10 points per leaf cell. If *numcells* $= 2$, then only the test cell and the nearest adjacent cell will be used in the search for nearest neighbors. So if *nn* > 20, then only the 20 points in these two cells will be returned. If *nn* is not specified (*n.s.*), then all points in all the *numcells* are used.

5. *C* : This parameter is called the *weighting* parameter. The weights are used in the least squares calculation of the derivatives. Once the nearest neighbors (called the learning points) have been determined, the normalized distance to the furthest point is known. The weight used for learning point *i* and test point *j* is computed as follows:

$$w_{ij} = exp(-kd_{ij}^2) \qquad (7.7.3)$$

where d_{ij} is the distance between the points and the value of *k* is computed based upon *C* as follows:

$$\frac{1}{C} = exp(-kd_{max}^2) \qquad (7.7.4)$$

where d_{max} is the distance to the furthest point. For example, if $C = 2$ then the weight for all points will be between 1 and 0.5. The smallest weight (i.e., 0.5) is given to the learning point that is farthest away from the test point.

The effect of *n* and **numcells** are shown in Table 7.7.1. The values of *runtime* are in CPU seconds measured using a Pentium II – 400 processor.

N	numcells	Nn	runtime
12500	All adj	100	9
25000	All adj	100	30
50000	All adj	100	109
12500	1	n.s.	2
25000	1	n.s.	6
50000	1	n.s.	27

Table 7.7.1   Timing results for several values of *n*. For all cases numleaves was 256 and *C* was 1. n.s. is "not specified".

In this table we see that the *runtime* increases at a rate that is much greater than linear. The *runtime* consists of two main components: the time to find the nearest neighbors and then the time to complete the least squares analysis. For the first 3 cases in which all adjacent cells were included in the search for the 100 nearest points, the number of points considered increases as $O(n)$ and thus the search for the 100 points increases as $O(n \, log_2 n)$. Since this search is repeated for all *n* points, the total time for this activity increases as $O(n^2 \, log_2 n)$. The least squares analysis for each point is the same since all 3 cases were based upon 100 points. However since this analysis is performed for each point, the total time for the least squares calculations is $O(n)$. The next three cases in the table are based upon using only the points in the test cells to perform the least squares analysis. Since *nn* is not specified, then all points in the cell are used. Thus each analysis is $O(n)$ and the total time for this activity is $O(n^2)$. However, since there is no need for a nearest neighbor search, the times for these 3 cases are considerably less than the times when all adjacent cells were used to find the 100 nearest neighbors.

In Table 7.7.2 we look more closely at the effect of **numcells**. At first glance the results look counter-intuitive. The average number of points in each cell is 50000 / 256 which is about 195. We see that going from 1 cell (i.e., the test cell) to 2 or 3 cells (i.e., the test cells plus the one or two closest cells) causes a large increase in the *runtime*. However, using all adjacent cells actually causes a slight decrease in *runtime*. When all adjacent cells are used the calculation required to find the closest cells is eliminated and this compensates for the increase in the number of points required to find the 100 nearest neighbors.

The next four cases use all points in the included cells to perform the least squares analyses. For **numcells** = 1, about 195 points are used, for **numcells** = 2, about 390 points are used until we reach a maximum when all points in all the adjacent cells are used. The last two cases show that most of the time spent (i.e., 335 − 39 = 296 seconds) was on the least squares analysis required to compute the derivatives.

Numcells	Nn	runtime	derivatives
1	100	26	Yes
2	100	112	Yes
3	100	112	Yes
All adj	100	109	Yes
1	n.s.	4	Yes
2	n.s.	54	Yes
3	n.s.	81	Yes
All adj	n.s.	335	Yes
All adj	n.s.	39	No

Table 7.7.2   Timing results for several values of *nuncells*. For all cases $n = 50000$, *numleaves* = 256 and $C = 1$. n.s. is "not specified".

Note the incredible speed attained when **numcells** = 1, $C = 1$, and **nn** is not specified. We call this *fast mode* because the points in the cell are only used once. The least squares calculation is performed using the absolute values of the x's and the u's rather than relative values the first time a test point lands in a cell and then the computed coefficients are saved. Thus a single complete second order polynomial is saved for each test cell and is used for all test points within the cell. Using this polynomial the first and second derivatives in all three directions can be computed for all points in the cell without repeating the least squares calculation. Of course, if new points are added to the cell, then the coefficients must be recomputed. The book-keeping required for this operation is included in the program code.

The effect of **numleaves** is seen in Table 7.7.3. In this table *preptime* (the time to create the p-tree needed for the nearest neighbor searching) is also included. By increasing the number of leaves, we see an increase in *preptime* (the time to build the tree) but this is insignificant when compared to the dramatic decrease in *runtime*. The decrease in *runtime* is due to the decrease in the total number of data points considered in the search for nearest neighbors. By increasing the number of leaf cells, we decrease the fraction of the total volume of the space used in the search. A word of caution should be added here. If we are looking for **nn** nearest neighbors, if *numleaves* is very large, the number of points per leaf cell might be so small that even if all adjacent leaves are considered, the total number of points examined will be less than *nn*. Clearly, as the number of points used in the least squares calculation decreases, the accuracy of the computed derivatives is adversely affected.

numleaves	nn	preptime	Runtime
256	100	2	109
512	100	3	72
1024	100	3	53
2048	100	4	33

Table 7.7.3   Timing results for several values of *nunleaves*. For all cases $n = 50000$, *numcells* = All adjacent cells and $C = 1$.

The effect of C is seen in Table 7.7.4. When $C > 1$ weights must be calculated (based upon distance from learning to the test points) and this explains the small increase in time required to compute derivatives ($80 - 10 = 70$ seconds for $C = 1$ and 74 seconds when $C > 1$). When $C > 1$ the distances to the test point must be computed because the distance to the furthest point is used in subsequent weight calculations. This additional calculation is performed regardless of whether or not derivatives are computed. The *runtime* increased from 10 to 21 seconds due to this added calculation.

C	preptime	runtime	derivatives
1.0	1	80	Yes
1.0	1	10	No
2.0	1	95	Yes
2.0	1	21	No

Table 7.7.4   Timing results for several values of C. For all cases $n = 25000$, *numleaves = 256, numcells* = All adjacent cells and *nn* is not specified.

In conclusion, this application illustrates the usage of kernel regression to obtain first and second order derivatives in three-dimensional space. To obtain realistic deformations, many atoms are required to describe the system. This is a very compute intensive application so a lot of attention to computational complexity is required. The key to reducing the time per calculation is in the reduction of the time required for nearest neighbor searching.

7.8 Applying Kernel Regression to Classification

Classification problems arise in many branches of science and technology. As an example, consider a production problem in which a simple test is required to decide whether or not a particular part should be accepted or rejected. Let us say that we have two very simple measurements that might be good predictors as to the quality of the part. Can we use these two measurements to decide in which class (accept or reject) a part falls?

In Section 2.8 application of least squares to classification problems was discussed. In general, for problems in which there are d independent variables, surfaces of d-1 dimensions were located to separate the classes. Thus if there are 2 independent variables and 2 classes, a single line is located to separate the classes. If there is only one independent variable, then the separation is accomplished with a single point. There are however, problems in which this approach cannot be used. An example of a distribution of two classes in a two dimensional space that cannot be separated in such a simple manner is seen in Figure 2.8.4.

Rather than trying to locate surfaces, the approach considered in this section is applicable to all types of distributions. As in Section 2.8, the classes are assigned numeric values. For example, for a two class problem, the values of $y = 0$ or 1 can be assigned to the data points from the two classes. The model is built using the *nlrn* points and tested using *ntst* test points. To predict the class of a test point, the methodology described in this chapter may be applied. For example, if we use the Order 0 algorithm, *nn* nearest neighbors from the *nlrn* set of points are first located and then a value of y is computed as described in Section 7.2. The learning points can be weighted according to Equation 7.2.1 based upon the distance to the test point or they can all be assumed to have equal weight. In addition, if the numbers of learning points from the classes is considerably different, the classes with lower numbers of cases can be assigned higher weights. This approach is discussed in Section 2.8 and is also applicable when kernel regression is used for classification. For two class problems if the computed value of y for the test point is less than 0.5 if would be predicted as a Class 0 point, otherwise it would be considered a Class 1 point. Alternatively, the δ technique (see Table 2.8.1) in which a "not clear" designation is assigned to some of the test points can be used. The use of Order 1 and 2 algorithms is certainly acceptable; however they should only be used if they outperform the simpler Order 0 algorithm.

For real problems in which there are a large number of available data points, one can experiment using various values of **nn** and perhaps using several different values of **C** (see Equation 7.6.3). The parameters could be selected based upon a criterion such as misclassification rate for the test data set. If there are enough data points available, it is useful to leave an evaluation data set untouched until the parameters have been set. This data set can be used to verify that the selected parameters lead to reasonable classifications for the as yet unseen data.

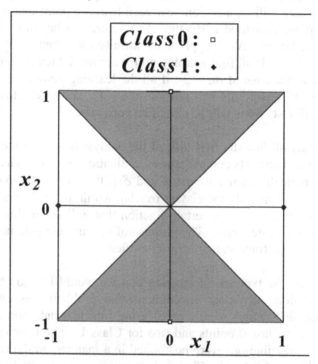

Figure 7.8.1   A Two-Class Two-dimensional Example

As an example, consider the two-class two-dimensional problem shown in Figure 7.8.1. In this example **nlrn** (the number of learning points) is 4 : 2 Class 0 points on the line $x_1 = 0$ and two Class 1 points on line $x_2 = 0$. Assume that the correct (but unknown model) is that all points in the shaded areas are from Class 0 and all points in the unshaded areas are from Class 1. The ranges of x_1 and x_2 are limited to -1 to 1. Using the Order Zero Algorithm and **nn** (the number of nearest neighbors) equal to one, all test points will be classified correctly. However, if we use **nn** = 2, for every test point we will locate one Class 1 learning point ($Y = 1$) and one Class 0 ($Y = 0$) learning point. If all points are weighted equally then the value of *ycalc* for every test point is 0.5. Points are classified as belonging to Class 0 if *ycalc* < 0.5 − δ, Class 0 if *ycalc* > 0.5 + δ and otherwise N.C. (not clear). Thus even if $\delta = 0$, all test points would fall into the N.C. category. If the 2 learning points are weighted according to Equation 7.6.3, for any value of $C > 1$ (i.e., $k > 0$) all points would be classified correctly.

If *nn* = 3, the situation becomes bizarre. If all points are weighted equally, then all test points will be misclassified! If the points are weighted using $C > 2$, then there will be a region near each learning point that test points in the region are classified correctly. As C increases beyond 2 these regions become larger. As C approaches infinity the situation approaches the case of *nn* = 1. If all four learning points are used, then the situation is very similar to the case of *nn* = 2 : if all the learning points are weighted equally, then all test points will fall into the N.C. category. However, if $C > 1$, then all test points will be classified correctly.

For a given set of data the first task of the analyst is to separate the data into learning and test sets and perhaps an evaluation data set. The parameters that are typically varied are *nn*, C and δ (if there is a need to create an N.C. category). Typically the classes overlap within the independent variable space so there will be a certain fraction that will fall within the misclassified or N.C. categories. The purpose of varying the parameters is to try to minimize the fraction that is misclassified.

To demonstrate the type of information that we would like to obtain, the following artificial two-class two-dimensional problem was developed. Ten thousand data points were created using four bivariate normal distributions: two for Class 0 points and two for Class 1. The locations of the centers of the distributions were positioned in a manner similar to the distribution in Figure 7.8.1. The Class 0 distributions were located on the line $x_1 = 0$ (at $x_2 = 3$ and −3). The Class 1 distributions were located on the line $x_2 = 0$ (at $x_1 = 3$ and −3). A value of $\sigma = 1$ was used in the bivariate distributions. Thus for the 2500 Class 1 points centered at $x_1 = 3$ and $x_2 = 0$, about 95% of the values of x_1 were in the range $3 − 1.96$ to $3 + 1.96$ and about 95% of the values of x_2 for this distribution were in the range −1.96 to 1.96. Results from a series of simulations are shown in Table 7.8.1.

nn	δ	Correct	Not Clear	Misclassified
1	0.0	0.9505	0.0000	0.0495
1	0.1	0.9505	0.0000	0.0495
1	0.2	0.9505	0.0000	0.0495
3	0.0	0.9610	0.0000	0.0390
3	0.1	0.9610	0.0000	0.0390
3	0.2	0.9140	0.0685	0.0175
10	0.0	0.9730	0.0000	0.0270
10	0.1	0.9650	0.0100	0.0250
10	0.2	0.9455	0.0355	0.0190
20	0.0	0.9705	0.0000	0.0295
20	0.1	0.9630	0.0145	0.0225
20	0.2	0.9500	0.0310	0.0190

Table 7.8.1   Correct, Not-Clear and Misclassified Rates as a Function of *nn* and δ For this example $C = 1$ (i.e., all learning points were equally weighted.

Once the data from the four distributions were combined, every fifth data point was put into the test data set and all the others were put into the learning data set. Thus the value of *nlrn* was 8000, and *ntst* was 2000. Results are shown in Table 7.8.1. We see that for this problem there is not a very dramatic difference as a function of *nn* but the results for *nn* = 10 are marginally best. For all values of *nn* the misclassification rate decreases as δ increases because some of the misclassified test points are transferred to the N.C. category. Also, some points that had been classified correctly are transferred to the N.C. category. Note that for *nn* = 1 the N.C. rate is 0 for all values of δ because the value of *ycalc* is either 0 or 1 depending upon the nearest neighbor.

The parameter *C* is used to set the value of *k* needed to calculate the weights of the learning points. From Equation 7.6.3 when $C = 1$ the value of $k = 0$ and therefore all points are equally weighted. When $C = 2$, the weight of the furthest point (from the *nn* nearest neighbors) gets a weight of 1/2 and all the other *nn* - 1 points get weights between 1/2 and 1. The effect of varying *C* is seen in Table 7.8.2. Note that the effect is not very dramatic for this particular problem. In fact, the results show a small advantage using $C = 1$ (i.e., weigh all points equally).

C	δ	Correct	Not Clear	Misclassified
1	0.0	0.9730	0.0000	0.0270
1	0.1	0.9650	0.0100	0.0250
1	0.2	0.9455	0.0355	0.0190
2	0.0	0.9690	0.0000	0.0310
2	0.1	0.9605	0.0130	0.0265
2	0.2	0.9440	0.0355	0.0205
3	0.0	0.9690	0.0000	0.0310
3	0.1	0.9605	0.0125	0.0270
3	0.2	0.9445	0.0345	0.0205
4	0.0	0.9695	0.0000	0.0305
4	0.1	0.9615	0.0110	0.0275
4	0.2	0.9445	0.0345	0.0210

Table 7.8.2   Correct, Not-Clear and Misclassified Rates as a Function of C and δ For this example $nn = 10$.

The order of the algorithm was also tested for this problem. Using Order 1 the results were close but not quite as good as results for Order 0. The best results were obtained using the simplest algorithm and simplest weighting scheme (i.e., all points are equally weighted). For problems in which the data density is high, the nearest neighbors are close to the test points and thus there is no particular advantage to going to higher order algorithms and decreasing weights for the further points. For this particular test problem, 8000 learning points were distributed in a two dimensional space and so in most regions of the space there were a sufficient number of nearest neighbors to make accurate classifications. However, for problems in higher dimensional spaces the data density decreases exponentially with the number of dimensions and one might expect that the higher order algorithms would be beneficial for such problems.

7.9 Group Separation: An Alternative to Classification

For some problems the analyst is interested in identifying groups with special properties. For example, an insurance company might want to develop a model for predicting whether or not a person applying for life insurance falls into a high risk or low risk group. A hedge fund operator might be looking for stocks which should outperform or under-perform the market. The traditional classification approach to problems of this type was discussed in Section 7.8. In this section the concept of "group separation" is introduced [WO00].

In Section 7.6 kernel regression performance studies were considered. The criterion for choosing parameters was maximization of VR (Variance Reduction). However, for some problems we are less interested in how a model performs for all the test points and are more interested in the performance of the top and bottom percentiles. For example, let us consider a model that uses kernel regression to predict the performance of stocks relative to the market. For a given date and stock the model yields a predicted relative return which can then be compared to the actual relative return. The question that the analyst is most interested in answering is whether or not the stocks with high predicted returns perform significantly better than stocks with low predicted returns. The group separation SEP is a criterion that may be used to answer this question:

$$SEP = \frac{Avg(G_1) - Avg(G_2)}{\sqrt{\sigma_{G1}^2 \Big/ n_1 - \sigma_{G2}^2 \Big/ n_2}} \qquad (7.9.1)$$

In this equation G_1 is the top group and G_2 is the bottom group. If Y is the actual variable that we are modeling, then for example, $Avg(G_1)$ is the average value of Y for group 1 and σ_{G1}^2 is the variance of Y for this group. Typically the values of n_1 and n_2 are the same (i.e., n) and are computed as a fraction of $ntst$ (the number of test points):

$$n = ntst * GroupPcnt / 100 \qquad (7.9.2)$$

If the model does not predict, then as the value of n becomes large we would expect SEP to be normally distributed with a mean of zero and a standard deviation of one. Thus SEP is a measure of separation of the groups that can easily be interpreted for significance.

For example, assume that we develop a model that predicts the relative one-day return of stocks and we test it over a period of 60 trading days. Further assume that the average number of stocks followed during this period is 8000 per day. We thus have 480 thousand data points that can each be compared to how the stocks actually performed for the days included in the study. Let us now question whether or not the top 5% of the stocks outperformed the bottom 5%. The value of n for each group is 24000. Let us assume that the mean and standard deviation for group 1 is 0.00037 ± 0.042 and for group 2 is -0.00024 ± 0.056. The value of **SEP** is:

$$SEP = \frac{0.00037 + 0.00024}{\sqrt{\dfrac{0.042^2 + 0.056^2}{24000}}} = 1.35$$

This result is not very significant. If the model predicts in a random manner, the probability of getting a value of 1.35 or greater is about 9%. For the same model, what happens when we limit the groups to 1%? Assume that the results for group 1 are 0.0070 ± 0.038 and for group 2 are -0.0088 ± 0.034. The value of **SEP** is:

$$SEP = \frac{0.0070 + 0.0088}{\sqrt{\dfrac{0.038^2 + 0.034^2}{4800}}} = 21.5$$

This number is extremely significant and suggests that there is a very large difference between the top and bottom 1% of the stocks when ranked on the basis of their predicted performance. For this example, the contrast between the top and bottom 1% as compared to the top and bottom 5% is quite striking!

Appendix A: Generating Random Noise

When evaluating any type of modeling software, it is useful to be able to create data sets that can be used for testing the software. Most general purpose statistical software packages contain random number generators. For example, MATLAB includes a function called **rand** that generates random numbers in the range 0 to 1 and another function called **randn** that generates random numbers from a normal distribution with a mean of zero and a standard deviation of one. These functions or their equivalents can be used to generate random noise satisfying the requirements of most data sets.

Let us assume **rand**(n) and **randn**(n) generate vectors of n random numbers. (The functions in MATLAB are quite general and can generate scalars, vectors or matrices.) Assume that we wish to create a Y vector from a vector $f(x)$ of n data points and we wish to add noise so that on average the noise component is 5% and the noise is normally distributed. The noise vector would be:

$$noise = 0.05 * f(x) * \mathbf{randn}(n) \tag{A.1}$$

and the Y vector would just be $f(x)$ + *noise*.

As a second example, assume that we wish to add noise to the $f(x)$ vector so that if the entire data set is modeled the noise is on average P times the actual signal. Assume that the noise is not a function of $f(x)$ and should be uniformly distributed within the range $-C$ to C. For this example, **rand** rather than **randn** is used. We must first generate a basic noise vector with a mean of zero:

$$noise = C \, (\, 2 \, \mathbf{rand}(n) - 1) \tag{A.2}$$

To compute C we note that the values of noise will be between $-C$ and C with a variance of $C^2/3$. It is reasonable to assume that the noise will be uncorrelated with the Y vector so the variance in the Y vector is:

$$\sigma_Y^2 = \sigma_f^2 + \sigma_{noise}^2 = \sigma_f^2 + C^2/3 \qquad (A.3)$$

If we want to create a data set in which the average noise is P times the signal (i.e., $f(x)$), we would first compute σ_f and then C as follows:

$$C = \sqrt{3P\sigma_f^2} = \sigma_f\sqrt{3P} \qquad (A.4)$$

For example, if we wish to create a data set that is 80% noise, then the value of P would be 4 and C would be $3.464\sigma_f$.

If we wish to repeat this example but prefer that the noise is generated using a normal distribution, Equation A.2 thru A.4 would be replaced by:

$$\boldsymbol{noise = C * \mathbf{randn}(n)} \qquad (A.5)$$

$$\sigma_Y^2 = \sigma_f^2 + \sigma_{noise}^2 = \sigma_f^2 + C^2 \qquad (A.6)$$

$$C = \sigma_f\sqrt{P} \qquad (A.7)$$

For the case of 80% noise, C would be $2\sigma_f$.

If there is no available equivalent to the **randn** function, **randn** can be generated from **rand**. The **randn** function generates random numbers from the standard normal u distribution. This distribution is tabulated in many sources (e.g., AB64, FR92, ST03). The tables include the probability of a point falling within the range from 0 to u. Theoretically u can range from $-\infty$ to $-\infty$, but for values of u above about 3 the probabilities are very close to 0.5. Using Equation A.8 (i.e., Equation 26.2.23 from AB64) n random probabilities in the range -0.5 to 0.5 are first generated: $p =$ **rand**$(n) - 0.5$. A vector of signs is then created: (**if** $p < 0$ $sign = -1$ **else** $sign = 1$). At this point only the absolute values of p are used ($p = \boldsymbol{abs}(p)$). Equation 26.2.23 computes the value of u_p such that the probability of a point falling above u_p is p. The normally distributed random numbers u_p corresponding to the random values of p are generated as follows:

$$t = \sqrt{ln\left(\frac{1}{p^2}\right)}$$

$$u_p = sign * \left(t - \frac{c_0 + c_1 t + c_2 t^2}{1 + d_1 t + d_2 t^2 + d_3 t^3} \right) + \varepsilon(p) \qquad \text{(A.8)}$$

where $\left|\varepsilon(p)\right| < 4.5 * 10^{-4}$. The values of the constants are:

$c_0 = 2.515517$ $d_1 = 1.432788$

$c_1 = 0.802853$ $d_2 = 0.189269$

$c_2 = 0.010328$ $d_3 = 0.001308$

For values of p approaching zero, t and therefore u_p becomes large. For example, for $p = 0.0001$, $t = 4.7985$ and $u_p = 4.2684$. For values of p approaching 0.5, u_p approaches zero. For example, for $p = 0.4999$, $t = 1.1776$ and $u_p = 0.00025$.

Appendix B: Approximating the Standard Normal Distribution

The standard normal distribution is probably the most widely used distribution in statistics. Indeed, many other distributions can be approximated by the standard normal as the number of events or data points becomes large. In classical statistics, the usage of the standard normal required the user to look up values in standard normal tables (e.g., AB64, FR92, ST03). To avoid the table lookup process, approximations to the standard normal are available and can be accessed as calls to functions from within software. General purpose statistical software packages include such functions.

The normal distribution was defined in Section 1.3 by Equation 1.3.7:

$$\Phi(x) = \frac{1}{\sigma(2\pi)^{1/2}} \exp(-\frac{(x-\mu)^2}{2\sigma^2}) \tag{1.3.7}$$

The standard normal distribution (denoted as the u distribution) is the normal distribution with a mean of zero and a standard deviation of one:

$$u(x) = \frac{1}{(2\pi)^{1/2}} \exp(-x^2 / 2) \tag{B.1}$$

The probabilities listed in the standard normal tables are the areas under the curve from 0 to u_p:

$$P(u_p) = \int_0^{u_p} u(x)\, dx \tag{B.2}$$

We can also define the $Q(u_p)$ as the area under the curve from u_p to ∞:

$$Q(u_p) = \int_{u_p}^{\infty} u(x)\,dx \qquad (B.3)$$

Note that $P(u_p) + Q(u_p) = 0.5$. In Appendix A, Equation A.8 is used to estimate the value of u_p for a given value of $Q(u_p)$. In this appendix, we consider the inverse problem: estimating either $Q(u_p)$ or $P(u_p)$ for a given value of u_p. In the Handbook of Mathematical Functions [AB64], Equation 26.2.17 can be modified to approximate $Q(u_p)$:

$$t = \frac{1}{1 + 0.2346419 * u_p}$$

$$z = exp(-u_p^2/2)$$

$$Q(u_p) = z * (b_1 t + b_2 t^2 + b_3 t^3 + b_4 t^4 + b_5 t^5) + \varepsilon(u_p) \qquad (B.4)$$

where $|\varepsilon(u_p)| < 7.5 * 10^{-8}$. The values of the constants are:

$$b_1 = 0.319381530 \qquad b_2 = -0.356563782$$
$$b_3 = 1.781477937 \qquad b_4 = -1.821255978$$
$$b_5 = 1.330274429$$

The equivalent value of $P(u_p)$ is just $0.5 - Q(u_p)$. As an example, for $u_p = 1.96$, the value of $Q(u_p)$ using B.4 is 0.024998 and $P(u_p)$ is 0.475002 which is in agreement with the value in the standard normal tables.

References

[**AB64**] M. Abramowitz, I. Stegen, *Handbook of Mathematical Functions*, NBS, 1964.

[**AR04**] Aronson, D. *Evidence Based Technical Analysis,* J. Wiley, 2006.

[**AZ94**] M. Azoff, *Neural Network Time Series Forecasting of Financial Markets*, Wiley, 1994.

[**BA74**] Y. Bard, *Nonlinear Parameter Estimation*, Academic Press, 1974.

[**Be61**] R.E. Bellman, *Adaptive Control Processes*, Princeton University Press, 1961.

[**BA94**] R. J. Bauer, *Genetic Algorithms and Investment Strategies*, Wiley, 1994.

[**BE03**] P. Bevington, D. Keith, *Data Reduction and Error Analysis for the Physical Sciences*, McGraw Hill, 2003.

[**CO99**] A. Constantinides, *Numerical Methods for Chemical Engineers with MATLAB Applications*, Prentice Hall, 1999.

[**DA90**] W.W. Daniel, *Applied Nonparametric Statistics, 2nd Edition*, PWS-Kent, 1990.

[**DA95**] M. Davidian, D. Giltinan, *Nonlinear Models for Repeated Measurement Data*, Chapman and Hall, 1995.

[**DE43**] W.E. Deming, *Statistical Adjustment of Data*, Wiley, 1943.

[**DR66**] N.R. Draper, H. Smith, *Applied Regression Analysis*, Wiley, 1966.

[**FR92**] J. Freund, *Mathematical Statistics (5th Edition)*, Prentice Hall, 1992.

[FO57] G.E. Forsythe, *Generation and Use of Orthogonal Polynomials for Data-Fitting with a Digital Computer*, J. Soc. Of Applied Math., Vol 5, No. 2, June 1957.

[GA57] C.F. Gauss, C.H. Davis, *Theory of the Motion of the Heavenly Bodies Moving about the Sun in Conic Sections: A Translation of Gauss's "Theoria Motus"*, Little Brown, 1857.

[GA84] T. Gasser, H. G. Muller, W. Kohler, L. Molianari, and A. Prader, *Nonparametric regression analysis of growth curves*, Annals of Statistics, 12, 210-229, 1984.

[GA94] P. Gans, *Data Fitting in the Chemical Sciences*, Wiley, 1992.

[GA95] E. Gately, *Neural Networks for Financial Forecasting*, Wiley, 1995.

[HA90] W. Hardle, *Applied Nonparametric Regression*, Cambridge University Press, 1990.

[HA01] T. Hastie, R. Tibshirani, J. Friedman, *The Elements of Statistical Learning*, Springer Series in Statistics, 2001.

[HU03] S. Huet, A. Bouvier, M. Poursat, E. Jolivet, *Statistical Tools for Nonlinear Regression. A Practical Guide with S-PLUS and R Examples*, Springer Series in Statistics, 2003.

[LE44] K. Levenberg, Quart. Applied Mathematics, Vol 2, 164, 1944.

[MA63] D.W. Marquardt, J. Soc. Ind. Applied Mathematics, Vol 11, 431, 1963.

[MC99] B. D. McCullough, *Econometric Software Reliability: EViews, LIMDEP, SHAZAM and TSP*, Journal of Applied Econometrics, Volume 14, Issue 2, 1999.

[ME77] M. Merriman, *The Elements of the Method of Least Squares*, J. Wiley & Sons, 1877.

[ME92] W. Mendenhall, T. Sincich, *Statistics for Engineers and Scientists*, 3rd Edition, Macmillan, 1992.

[MO60] R. H. Moore, R. K. Zeigler, *The Solution of the General Least Squares Problem with Special Reference to High Speed Computers*, LA-2367 Report, March 1960.

[MO98] D. Mount, S. Arya, *ANN: Library for Approximate Nearest Neighbor Searching*, University of Maryland, 1998.

[MO03] H. Motulsky, A. Christopoulos, *GraphPad Prism – Fitting Models To Biological Data using Linear and Nonlinear Regression*, GraphPad Software, 2003.

[PY99] D. Pyle, *Data Preparation for Data Mining*, Morgan Kaufmann, 1999.

[RA73] C. R. Rao, *Linear Statistical Inference and its Applications, 2nd Edition*, Wiley, 1973.

[RA78] A. Ralston, P. Rabinowitz, *A First Course in Numerical Analysis*, McGraw Hill, 1978.

[RE95] A.P. Refenes, *Neural Networks in the Capital Markets*, Wiley, 1995.

[SA90] H. Samet, *The Design and Analysis of Spatial Data Structures*, Addison Wesley Longman, 1990.

[SC98] Oliver Schabenberger, *Nonlinear Regression with the SAS System*, http://www.ats.ucla.edu/stat/sas/library/SASNLin_os.htm, 1998.

[SH48] C. E. Shannon, *A Mathematical Theory of Communication*. The Bell System Technical Journal, Vol 27, 379-423, 623-656, July, Oct 1948. also http://cm.bell-labs.com/cm/ms/what/shannonday/shannon1948.pdf.

[SI88] S. Siegel, N. J. Castellan, *Nonparametric Statistics for the Behavioral Sciences*, McGraw Hill, 1988.

[SK97] Skiena, S. S. *Nearest Neighbor Search*. §8.6.5 in *The Algorithm Design Manual*. Springer-Verlag, pp. 361-363, 1997.

[SM00] Smid, M. "Closest-Point Problems in Computational Geometry." Ch. 20 in *Handbook of Computational Geometry*. North-Holland, pp. 877-935, 2000.

[ST86] S. M. Stigler, *The History of Statistics: the Measurement of Uncertainty before 1900*, Harvard University Press, 1986.

[ST03] *Statistica Electronic Textbook,* www.statsoft.com/textbook/ stathome.html, Distribution Tables, StatSoft Inc, 2003.

[TA03] E. Tadmor, A. Agrawal, J. Wolberg, *An Efficient Data Structure for Graded Atomistic Simulations,* Proc of the 29th Israel Conf. of Mechanical Eng, May 2003

[TV04] J. Tvrdik, I. Krivy, *Comparison of Algorithms for Nonlinear Regression Estimates,* Compstat'2004 Symposium, Physcia-Verlag/Springer 2004.

[UL93] A. Ullah and H. D. Vinod, *General Nonparametric Regression Estimation and Testing in Econometrics,* Chapter 4, Handbook of Statistics 11, North Holland, 1993.

[VE02] W. Venables, B. Ripley, *Modern Applied Statistics with S,* 4th Edition, Springer-Verlag, 2002.

[WA93] R. E. Walpole, R. H. Meyers, *Probability and Statistics for Engineers and Scientists, (5th edition),* McMillan, 1993.

[WI62] Wilks, S. S., *Mathematical Statistics,* Wiley, 1962.

[WO62] J. R. Wolberg, T.J. Thompson, I. Kaplan, *A Study of the Fast Fission Effect in Lattices of Uranium Rods in Heavy Water,* M.I.T. Nuclear Engineering Report 9661, February, 1962.

[WO67] J. R. Wolberg, *Prediction Analysis,* Van Nostrand, 1967.

[WO71] J. R. Wolberg, *Application of Computers to Engineering Analysis,* McGraw Hill, 1971.

[WO00] J. R. Wolberg, *Expert Trading Systems: Modeling Financial Markets with Kernel Regression,* Wiley, 2000.

[ZE98] J. Zelis, *SPSS Advanced Statistics 6.1,* UCL Institut de Statistique, 1998.

Index